ELEMEN
VECTOR A.

BY

L. SILBERSTEIN, Ph.D.

LECTURER IN NATURAL PHILOSOPHY AT THE UNIVERSITY OF ROME

PREFACE

THIS little book was written at the instance of Messrs. Adam Hilger, and, in accordance with their desire, it contains just what is required for the purpose of reading and handling my *Simplified Method of Tracing Rays, etc.* (Longmans, Green & Co., London, 1918). With this practical aim in view, all critical subtleties have been purposely avoided. In fact, it is scarcely more than a synoptical presentation of the elements of Vector Algebra covering the needs of those engaged in geometrical optics. At the same time, however, it is hoped that this booklet will serve a more general purpose, viz. to provide everybody unacquainted with the subject with an easy introduction to the use of Vector Algebra.

It is scarcely necessary to explain that the deductions given in this book are based on Euclid's axioms, notably with the inclusion of his postulate of parallels—upon which the equality of vectors is most essentially based. Those readers who are desirous of seeing how the formal rules here given can be generalized so as to be valid independently of the axioms of congruence and of parallels, may consult the author's *Projective Vector Algebra* (Bell & Sons, 1919), and a sequel to it published in *Phil. Mag.* for July, 1919, pp. 115-143. It is, however, advisable for the student to become first thoroughly familiar with the euclidean vector algebra as here presented.

I take the opportunity of expressing my sincere thanks to Messrs. Hilger for enabling me to make this further contribution towards the promotion of the more general use of this powerful and convenient language of vectors, and to the Publishers for the care they have bestowed upon this little book. L. S.

CONTENTS

PAGE

1. VECTORS DEFINED - - - - - - - - 1
2. EQUALITY OF VECTORS DEFINED - - - - - 2
3. ADDITION OF VECTORS - - - - - - - - 3
4. SUBTRACTION OF VECTORS - - - - - - - 10
5. SCALAR PRODUCT OF TWO VECTORS - - - - - 11
6. THE VECTOR PRODUCT OF VECTORS - - - - - 17
7. EXPANSION OF VECTOR FORMULAE - - - - - 21
8. ITERATION OF VECTORIAL MULTIPLICATION - - - 23
9. THE LINEAR VECTOR OPERATOR - - - - - - 25
10. HINTS ON DIFFERENTIATION OF VECTORS - - - - 38

INDEX - - - - - - - - - - - 41

ELEMENTS OF VECTOR ALGEBRA

1. Vectors defined. Whereas common algebraic magnitudes, such as the number of inhabitants of a village, or the mass of a body, or the energy stored in an accumulator, having nothing to do with direction, are called *scalars*, any magnitude such as a displacement, a velocity or an acceleration, which has size as well as *direction* in space, is called a *vector*. The visual, or tangible, representative of any vector whatever is a segment of a straight line of some length, representing the vector's size, and of some definite direction in space, together with its *sense* (say, from a point M towards a point N), giving the direction of the vector.

Vectors will be printed in Clarendon, thus

$$\mathbf{A},\ \mathbf{B},\ \text{etc.,} \quad \text{or} \quad \mathbf{n},\ \mathbf{r},\ \mathbf{s},\ \text{etc.,}$$

and their sizes, regardless of direction, or their *tensors* (as they are called) will be denoted by the same letters in Italics. Thus, A will be the tensor of $\mathbf{A}$; B, n will be the tensors of $\mathbf{B}$, $\mathbf{n}$, and so on.

Returning once more to the above definition, we may as well say that any vector $\mathbf{A} = OE$ is given by the ordered couple or *pair* of points, O the *origin* and E the end-point of the vector; the tensor, called also the absolute value, of the vector being the mutual distance of O and E. In short symbols, and using the familiar bar for the distance,

$$\mathbf{A} = O \rightarrow E, \quad A = \overline{OE}.$$

The tensor of a vector is thus an ordinary, absolute or essentially positive number.

A vector whose tensor is (in a conventionally fixed scale) equal to unity, is termed an *unit vector*. Thus, if $r = 1$, the corresponding $\mathbf{r}$ will be a unit vector. It will be understood that the denomination of A is that of $\mathbf{A}$. That is to say, if $\mathbf{A}$ is, for instance, the displacement of a particle, A will mean so many centimetres; and if **A** represents a velocity, A will be a number of cm. per second, and so on.

As far as will be possible we shall reserve small (in distinction from capital) Clarendon letters for unit vectors. Thus, if the contrary is not expressly stated, **a**, **b**, etc., will stand for unit vectors, so that $a=1$, $b=1$, etc.

In MS. work the reader will, at least in the beginning of his vector career, find it useful to underline all his vectors once or twice. Or he may write them thicker, and imitating somehow the printer's type. Then, everyone will soon find out his most agreeable manner of writing.

2. Equality of vectors defined. We have just seen that the two essential features of a vector are: its size or tensor, and its direction in space.

In some branches of physico-mathematics it is important to consider the position of the vectors in question (besides their sizes and directions), *i.e.* to localize their origins, either by fixing the origin of each vector altogether or by allowing it only to move freely in its own line. Such vectors are usually called "localized" vectors. In a vast class of investigations, however, the position of these directed magnitudes is of no avail, and it is then obviously convenient *not* to include position among the determining characteristics of a vector. Such vectors, in distinction from localized ones, are called *free* vectors. These and these only will here occupy our attention. The adjective will be dropped, however, and the beings in question will be called shortly vectors. With this understanding, the definition of their equality may be put thus:

By saying that two vectors, **A** and **B**, are *equal to one another*, and by writing

$$\mathbf{A}=\mathbf{B} \quad \text{or} \quad \mathbf{B}=\mathbf{A},$$

we mean that their *tensors are equal*, $A=B$, and that they have *the same direction* or, in other words, that the straight segments representing these two vectors have *the same length* and are concurrently *parallel* to one another. In short symbols, $\mathbf{A}=\mathbf{B}$ means as much as

$$A=B \quad \text{and} \quad \mathbf{A}\uparrow\uparrow\mathbf{B}.$$

Thus, if a pair of points, O, E, represents a vector $\mathbf{A}=O\rightarrow E$, the ∞^3 point pairs O', E' or straight segments $O'E'$ of equal length with and concurrently parallel to OE are all equal to **A**, no matter where their origins are situated. Notice that through every point O' of euclidean space there is one and only one parallel to OE, so that from every space point O' as origin one and only one vector can be drawn which is equal to the given **A**. Of course, the laying off, from O', of the length $\overline{O'E'}=\overline{OE}$ implies the use of some " rigid transferer," such as a pair of compasses.

Equivalently, we may say that the rigid *translation* (parallel shifting) of a given vector is irrelevant, or does not change the vector. Provided it is not being rotated, stretched or contracted, we can, by the accepted definition, " transfer " it to any place we like best.

Two vectors **A**, **B** drawn from the same origin are termed *coinitial.* By what has just been said, *any* two vectors can be made coinitial, by shifting one of them or both parallel to themselves. If $\mathbf{A}=\mathbf{B}$, then making them coinitial, fuses them into one straight segment. If only $A=B$ (equal tensors only), then making the vectors **A**, **B** coinitial will still leave a certain non-vanishing angle, or direction difference, between them, sufficient by itself to declare the two vectors as being different from one another.

We will say that two or more vectors form a *chain* if the end-point of one serves as the origin for the other, and so on. As before, *any* two vectors **A**, **B** can be linked up into a chain, to wit in two manners: end-point of **A** coinciding with origin of **B** (or **A** preceding **B**), or *vice versa.*

This licence will be seen to be of capital importance for the vector sum to be defined presently, inasmuch as it will confer upon that sum the extremely convenient property of commutativity. It will, therefore, be important to keep these latter, apparently trivial remarks well in mind.

3. Addition of vectors. Let **A** and **B** be any two vectors, drawn anywhere. Shift **B** so as to bring its origin to coincidence with the end-point of **A**, as shown in Fig. 1. The vectors being thus linked up into a chain we call *sum* of **A** and **B** and denote by

$$\mathbf{S}=\mathbf{A}+\mathbf{B}$$

a third vector **S** which runs *from the beginning to the end of the chain, i.e.* from the origin of **A** to the end-point of **B**.

This is the definition of the vector sum. The operation, vector *addition*, thus defined has the so-called group-property, that is to say, being performed on vectors it gives again a vector, in much the same way as five apples added to three apples give again a certain number of apples.

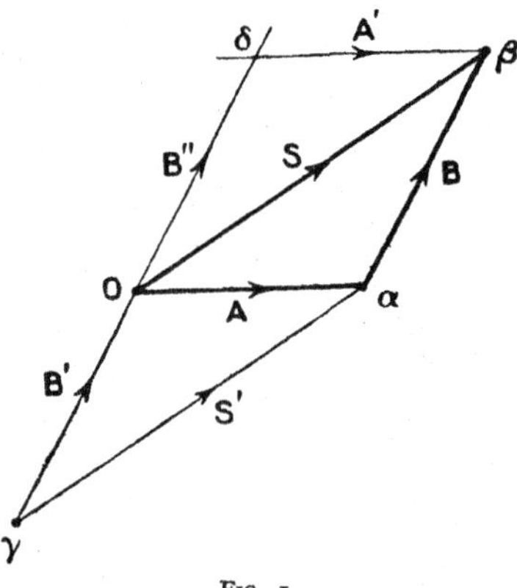

FIG. 1.

The above vectorial expression will be read: **B** added to **A.** But we might as well have linked the two given vectors so that the end-point of $\mathbf{B}=\mathbf{B}'$ falls into the origin of **A**, as shown in the lower part of Fig. 1. Then their sum, say $\mathbf{S}'$, would—according to the definition—be

$$\mathbf{S}'=\mathbf{B}+\mathbf{A},$$

which reads: **A** added to **B**. The natural question arises: What is this new vector $\mathbf{S}'$? Is it equal to **S**?

The answer is in the affirmative. For, by construction, $\mathbf{B}'$ is parallel to **B** and $B'=B$, so that $O\alpha\beta$ and $\alpha O\gamma$ are congruent triangles, and $S'=S$. At the same time the angles β and γ are equal to one another, so that, $\alpha\beta$ being parallel to γO, so are also $O\beta$ and $\gamma\alpha$, or $\mathbf{S}\uparrow\uparrow\mathbf{S}'$. Therefore, by Section 2, $\mathbf{S}'=\mathbf{S}$, what was to be proved.

Thus we have

$$\mathbf{A}+\mathbf{B}=\mathbf{B}+\mathbf{A}, \tag{1}$$

the *commutative* property of vector addition. The order of the addends, in the vector chain, is irrelevant for their sum.

Again, we might have shifted **B** to the position $\mathbf{B}''$ (Fig. 1), retaining also the previous $\mathbf{B}=\alpha\rightarrow\beta$ and constructing $\mathbf{A}'=\delta\rightarrow\beta=\mathbf{A}$.

Then, $O\alpha\beta\delta$ being a parallelogram and $\mathbf{S}=O\rightarrow\beta$ one of its diagonals, we should have the following construction of the sum of two coinitial vectors **A**, **B** (Fig. 2):

Through α, the end-point of **A**, draw a parallel to **B**, and through β, the end-point of **B**, draw a parallel to **A**. Then γ, the cross of

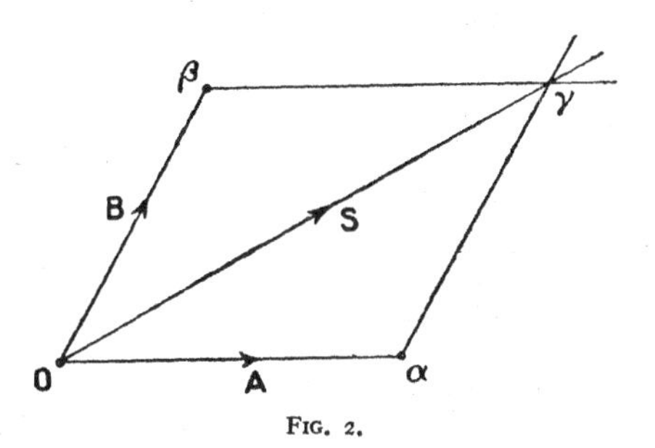

FIG. 2.

these parallels, will be the end-point of the required vector sum $\mathbf{A}+\mathbf{B}$ or $\mathbf{B}+\mathbf{A}$, and the common origin O of the two addends will be the origin of their sum

$$\mathbf{S}=O\rightarrow\gamma=\mathbf{A}+\mathbf{B}=\mathbf{B}+\mathbf{A}. \tag{2}$$

This is known as the parallelogram construction of a vector sum.

We might have started from it as a sum definition. It has the advantage of being immediately *symmetrical* with respect to the two addends. At any rate we see that the chain and the parallelogram constructions are (in virtue of Euclid) wholly equivalent to one another.

Thus far the case of two vector addends. Now, the sum of these being again a vector, $\mathbf{S}=\mathbf{A}+\mathbf{B}$, we can add to **S** any third vector **C**, thus obtaining

$$\mathbf{S}+\mathbf{C}=(\mathbf{A}+\mathbf{B})+\mathbf{C}=\mathbf{C}+(\mathbf{A}+\mathbf{B}),$$

the latter by the commutative property. Similarly for the sum of four and more vectors. Again, linking up the vector addends **A**, **B**, **C** into a chain, we see without difficulty (Fig. 3) that

$$(\mathbf{A}+\mathbf{B})+\mathbf{C}=\mathbf{A}+(\mathbf{B}+\mathbf{C}), \tag{3}$$

the result being in both cases the same vector, viz. that drawn from the beginning to the end of the chain. The same property holds for the sum of any number of vectors. The brackets become superfluous, and either of the above expressions can simply be written

$$\mathbf{A}+\mathbf{B}+\mathbf{C}$$

or $\mathbf{B}+\mathbf{A}+\mathbf{C}$, and so on.

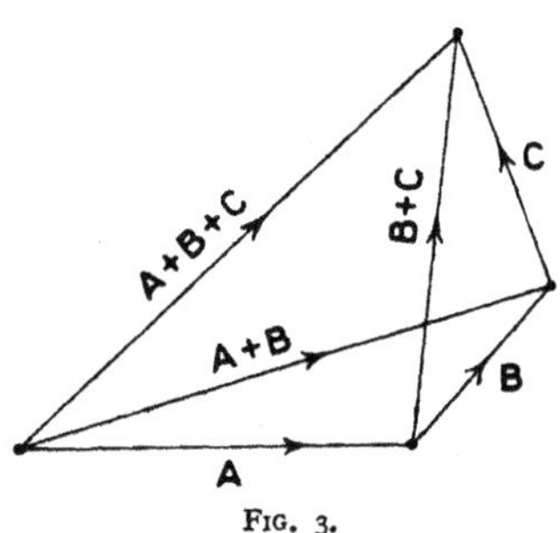

FIG. 3.

The addition of vectors is thus seen to be *associative* as well as *commutative*, exactly as the ordinary algebraic addition of scalars.

If by any appropriate parallel shifting of any number of given vectors, say **A**, **B**, **C**, **D**, they can be linked up, as in Fig. 4, into

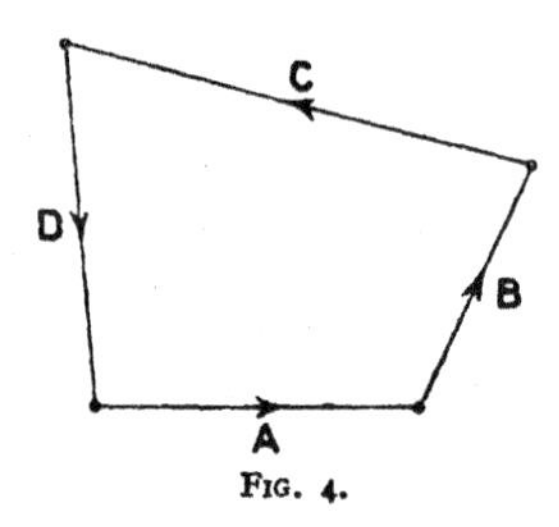

FIG. 4.

a *closed* chain (or a polygon), plane or not, then the sum of these vectors is a *nil vector* or simply *nil*,

$$\mathbf{S}=\mathbf{A}+\mathbf{B}+\mathbf{C}+\mathbf{D}=0,$$

and therefore also $\mathbf{A}+\mathbf{C}+\mathbf{B}+\mathbf{D}=0$, etc. It is scarcely necessary to say that a vector is *nil* or *zero*, $\mathbf{S}=0$, if

$$S=0,$$

that is, if its tensor vanishes, and conversely; or, in other words, if its end-point and origin coincide, such precisely being the case of our closed chain.

The vector sum, which shares with the ordinary algebraic sum the two capital properties of commutativity and associativity, contains the algebraic sum as a particular sub-case, to wit, when the vector addends are all parallel to one another. For, such being the case, they can always be brought into one line or made collinear. Parallel vectors, no matter what their tensors, are therefore called also *collinear vectors*. Now, if **A**, **B** are collinear vectors, the tensor of their sum is

$$A \pm B,$$

according as **A**, **B** are of equal or of opposite senses.

The tensor of a sum of vectors, as $\mathbf{S}=\mathbf{A}+\mathbf{B}$, can conveniently be denoted by

$$S=|\mathbf{A}+\mathbf{B}|,$$

as is usual for the absolute value of ordinary algebraic magnitudes. Thus we shall have, for *collinear* vectors,

$$|\mathbf{A}+\mathbf{B}|=|A \pm B|.$$

But, it will be well kept in mind that in general, for non-collinear addends,

$$|\mathbf{A}+\mathbf{B}| \neq |A \pm B|,$$

since $|\mathbf{A}+\mathbf{B}|$ is the length of the third side of a triangle whose two other sides are A and B.

By what has just been said, the sum of two equal vectors, which is written

$$\mathbf{A}+\mathbf{A} \quad \text{or} \quad 2\mathbf{A},$$

is a vector coinciding with **A** in direction and having $2A$ for its tensor. Similarly for $3\mathbf{A}$, $4\mathbf{A}$, and so on.

Again, if **B** be such a vector that

$$2\mathbf{B}=\mathbf{A},$$

we shall write

$$\mathbf{B}=\tfrac{1}{2}\mathbf{A},$$

and similar meanings will be attached to $\tfrac{1}{3}\mathbf{A}$, $\tfrac{1}{4}\mathbf{A}$, etc. In this manner, and applying in the case of irrational factors the well-known limit-reasoning, we easily obtain the meaning of the expression

$$n\mathbf{A},$$

where n is any positive scalar number, integral, fractional or irrational. We can say shortly that $n\mathbf{A}$ is the vector **A** stretched in the ratio $n:1$. If n is negative, then (as justified in Section 4, *infra*) $n\mathbf{A}$ will be the vector $\mathbf{A}$ stretched in the ratio $|n| : 1$ and then *inverted* in its sense, or first inverted and then stretched.

In particular, if $\mathbf{a}$ is an unit vector, the "unit of $\mathbf{A}$," as we have said before, we shall obviously have

$$\mathbf{A} = A\mathbf{a}. \tag{4}$$

Here A, the tensor of $\mathbf{A}$, is an ordinary positive number.

Let $\mathbf{a}$, $\mathbf{b}$ be any two *non-collinear* unit vectors. (Imagine them shifted so as to be coinitial.) Then any vector $\mathbf{R}$ contained in or parallel to the plane $\mathbf{a}$, $\mathbf{b}$ can obviously be expressed by

$$\mathbf{R} = x\mathbf{a} + y\mathbf{b}, \tag{5}$$

where x, y are some scalar numbers. For the plain meaning of this assertion is that, starting from O, the origin of $\mathbf{a}$, $\mathbf{b}$, any other point of the plane $\mathbf{a}$, $\mathbf{b}$ can be reached by making a number (x)

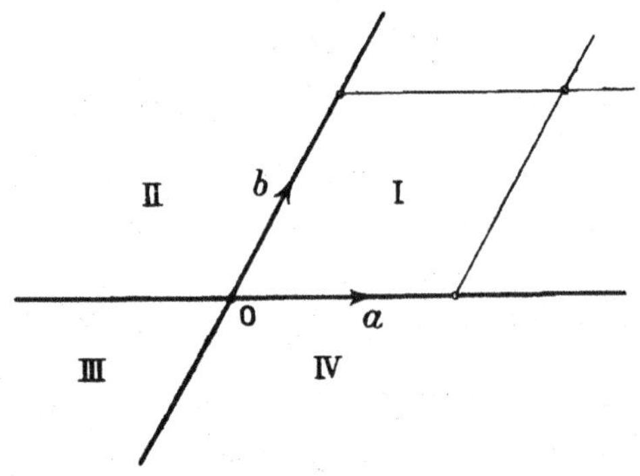

FIG. 5.

of steps $\mathbf{a}$ and then a number (y) of steps $\mathbf{b}$ (or first $y\mathbf{b}$ and then $x\mathbf{a}$). If both x and y are positive, then, with O as origin, $\mathbf{R}$ will lie in the region **I** of the plane $\mathbf{a}$, $\mathbf{b}$ (Fig. 5); if $x < 0$, $y > 0$, it will fall into II; if $x < 0$, $y < 0$, into III, and finally, if $x > 0$, $y < 0$, into the region IV.

The scalars x, y in (5) are called *the components of* $\mathbf{R}$ *along* $\mathbf{a}$, $\mathbf{b}$ *as axes.*

Similarly, if $\mathbf{a}$, $\mathbf{b}$, $\mathbf{c}$ be any three *non-coplanar* vectors * which we may again take as unit vectors, then any vector whatever can be expressed in the form

$$\mathbf{R} = x\mathbf{a} + y\mathbf{b} + z\mathbf{c}. \tag{6}$$

* *I.e.* such as cannot be made coplanar by parallel translations.

The scalars x, y, z are called the components of $\mathbf{R}$ taken along $\mathbf{a}$, $\mathbf{b}$, $\mathbf{c}$ as axes. These axes may be chosen at our will (if we wish at all to split our vectors $\mathbf{R}$ into components), either perpendicularly or obliquely to one another, the only condition for covering all possible vectors ($\mathbf{R}$) being that $\mathbf{a}$, $\mathbf{b}$, $\mathbf{c}$ should not be coplanar.

The three vectors $\mathbf{a}$, $\mathbf{b}$, $\mathbf{c}$ or, as we will say, the *reference system* $\mathbf{a}$, $\mathbf{b}$, $\mathbf{c}$, being fixed conventionally, we see from (6) that any vector is fully determined by *three* scalar data x, y, z and not less than three. The same thing is obvious from formula (4), according to which any vector $\mathbf{R}$ can be represented by

$$R\mathbf{r}.$$

In fact R is one scalar number, and $\mathbf{r}$ (a direction) implies two more scalar data, for instance two angles, which makes in all *three* independent scalar data as above.

In so-called polar coordinates, for instance, we have (Fig. 6)

$$\mathbf{R}=R\{[\mathbf{i}\cos\phi+\mathbf{j}\sin\phi]\sin\theta+\mathbf{k}\cos\theta\}, \tag{7}$$

where $\mathbf{i}$, $\mathbf{j}$, $\mathbf{k}$ are mutually perpendicular unit vectors, the pole-distance or co-latitude θ being counted from $\mathbf{k}$ and the longitude from $\mathbf{i}$ towards $\mathbf{j}$, if $\mathbf{i}$, $\mathbf{j}$, $\mathbf{k}$ be a right-handed system. The particular

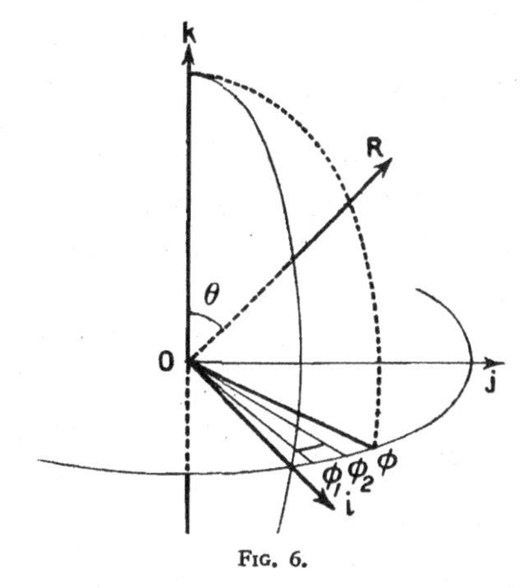

FIG. 6.

form (7) will often be found useful in passing from vector to scalar formulae, especially in optical computations. The unit of $\mathbf{R}$, *i.e.* any unit vector $\mathbf{r}$, will be, in polar coordinates,

$$\mathbf{r}=[\mathbf{i}\cos\phi+\mathbf{j}\sin\phi]\sin\theta+\mathbf{k}\cos\theta. \tag{7a}$$

After what has been said, it is scarcely necessary to explain that every vector equation is equivalent to *three* scalar ones. For to give **R** is equivalent to giving, say, its three rectangular or "cartesian" components, or, as in (7), the polar coordinates R, θ, ϕ of the end-point of **R**, with O as origin. Thus **A** = **B** means as much as $A_1 = B_1$ and $A_2 = B_2$ and $A_3 = B_3$, if the suffixes 1, 2, 3 are used for the components of the vectors along **i**, **j**, **k**, or as much as

$$R_A = R_B, \quad \theta_A = \theta_B, \quad \phi_A = \phi_B,$$

if the suffixes $_{A,\,B}$ are used to distinguish the polar coordinates of the end-point of **A** from those of the end-point of **B**.

But it must henceforth be urged that any such splitting of a vector should be avoided as much as is possible in the course of a vector investigation of any kind. For the utility of vector method lies precisely therein that it enables us to treat vectors as wholes instead of the triad of "components" of each of them.

4. Subtraction of vectors. This will require but a few remarks. In fact, as in common algebra, *the difference* of two vectors **A** and **B**, to be denoted by

$$\mathbf{A} - \mathbf{B},$$

may be defined as such a vector **C**, which added to **B** gives **A**. In symbols, we say that

$$\mathbf{C} = \mathbf{A} - \mathbf{B},$$

if

$$\mathbf{B} + \mathbf{C} = \mathbf{A}. \quad (8)$$

From this definition we see at once that if **A**, **B** are made coinitial (Fig. 7), the vector **A** − **B** *runs from the end-point of* **B** *to the end-point of* **A**. From the same figure, and by what was explained

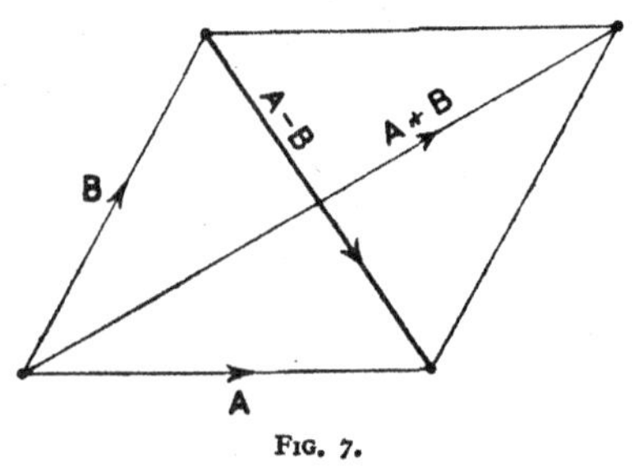

Fig. 7.

previously, we see that **A** + **B** and **A** − **B** are represented by the two diagonals of the parallelogram constructed upon **A**, **B**.

Apply the above definition (8) to the particular case $\mathbf{A}=0$; then

$$\mathbf{C}=0-\mathbf{B}=-\mathbf{B},$$

and $\mathbf{B}+\mathbf{C}=0$; therefore,

$$\mathbf{B}+(-\mathbf{B})=0. \tag{9}$$

This settles the meaning of the vector denoted by $-\mathbf{B}$; it is the vector which runs from the end-point towards the origin of **B**, or the reverse of **B**. This also justifies the interpretation given before to a negative scalar factor of a vector. Henceforth, for any **A**, **B**,

$$\mathbf{A}-\mathbf{B}$$

will stand for the same vector as

$$\mathbf{A}+(-\mathbf{B}).$$

The above remarks complete the meaning of

$$n\mathbf{A},$$

where **A** is a vector and n *any* real scalar, positive, nil or negative. The concept of such a product of a vector by any *scalar* n does not contain, in fact, anything besides the previous concept of vector sum or difference. It is derived from their special case, viz. relating to collinear vectors.

To say it once more, $n\mathbf{A}$ is simply the vector **A** stretched in the ratio $|n|:1$ and, if $n<0$, turned through $180°$ (in any plane passing through **A**).

Finally, as the reader himself will easily prove, for any **A**, **B**, and any scalar factor n,

$$n(\mathbf{A}+\mathbf{B})=n\mathbf{A}+n\mathbf{B}. \tag{10}$$

Similarly for three or more vector addends. This settles all questions concerning the multiplication (or division) of a vector expression by any scalar number.

5. Scalar product of two vectors. We now come to a new concept, transcending that of vector addition which hitherto has occupied us. The "scalar product" of two vectors **A**, **B** which will be denoted by **AB** is, first of all, not a vector but a *scalar*. (Thus the scalar multiplication of vectors does not respect the group requirement; it yields a result not contained in the class of operands: it takes two vectors and constructs out of them something which is utterly deprived of direction. None the less it is a very useful operation.) The value of this scalar is, by definition, proportional to the tensors of both the factors and to the *cosine* of the angle (**A**, **B**) included between them. In short, the definition of the *scalar product* is

$$\mathbf{AB} = AB \cos(\mathbf{A}, \mathbf{B}). \tag{11}$$

This can also be read: **AB** is the projection of **A** upon **B** multiplied by B, or the projection of **B** upon **A** multiplied by A.

Since $AB = BA$, for A, B are common numbers, and

$$\cos(\mathbf{A}, \mathbf{B}) = \cos(\mathbf{B}, \mathbf{A}),$$

we see at once from the very definition (11) that

$$\mathbf{AB} = \mathbf{BA}, \tag{12}$$

the *commutative* property.

According as the angle (**A**, **B**) is $< \frac{\pi}{2}$ or $> \frac{\pi}{2}$ $\left(\text{but } < \frac{3\pi}{2}\right)$, the product **AB** is positive or negative; for A, B are themselves essentially positive. And if $(\mathbf{A}, \mathbf{B}) = \frac{\pi}{2}$ or $\mathbf{A} \perp \mathbf{B}$, then

$$\mathbf{AB} = 0,$$

no matter what the (finite) tensors of **A**, **B**. In this case the operation (scalar multiplication) deprives the material operated upon not only of direction but of size. It annihilates it.

Conversely, if of two vectors **A** and **B** we know only that $\mathbf{AB} = 0$, then the only conclusion we can draw from it is that

$$\mathbf{A} \perp \mathbf{B},$$

but by no means that one of the factors vanishes, unless we happen to know beforehand that the two vectors cannot be perpendicular. It is of prime importance to keep this well in mind:

$\mathbf{AB} = 0$ means in general only as much as $\mathbf{A} \perp \mathbf{B}$.

The scalar product **AB** contains the ordinary algebraic product as a special case, to wit, when **A**, **B** are *collinear* vectors. For if such be the case, we have $\cos(\mathbf{A}, \mathbf{B}) = \pm 1$, and therefore,

$$\mathbf{AB} = \pm AB, \tag{13}$$

according as **A** and **B** have the same or opposite directions.

Since the tensor of the vector $m\mathbf{A}$ is mA, we see at once from (11) that

$$m\mathbf{A}n\mathbf{B} = mn\mathbf{AB}.$$

Thus, for example, if $\mathbf{a}$, $\mathbf{b}$ be the units of $\mathbf{A}$, $\mathbf{B}$, we have

$$\mathbf{AB} = AB\mathbf{ab},$$

where, again by the definition (11),

$$\mathbf{ab} = \cos(\mathbf{a}, \mathbf{b}), \tag{14}$$

valid for any pair of *unit* vectors $\mathbf{a}$, $\mathbf{b}$. Thus, for instance, if $\mathbf{a}$, $\mathbf{b}$ make with one another the angle of 45°, we have $\mathbf{ab} = \frac{1}{\sqrt{2}}$, and if $(\mathbf{a}, \mathbf{b}) = 90°$, $\mathbf{ab} = 0$. For the three normal unit vectors $\mathbf{i}$, $\mathbf{j}$, $\mathbf{k}$ used above we have $\mathbf{ij} = \mathbf{jk} = \mathbf{ki} = 0$.

As a sub-case of (13) we have the scalar *square* of a vector, or better, its *autoproduct*,

$$\mathbf{AA} \quad \text{or} \quad \mathbf{A}^2 = A^2,$$

and if $\mathbf{a}$ be a *unit* vector,

$$\mathbf{a}^2 = a^2 = 1.$$

Thus, $\mathbf{i}^2 = \mathbf{j}^2 = \mathbf{k}^2 = 1$.

Again, if $\mathbf{R}$ is any vector whatever and $\mathbf{n}$ a unit vector, $\mathbf{Rn}$ is the (scalar) component of $\mathbf{R}$ along $\mathbf{n}$, or the orthogonal projection of $\mathbf{R}$ upon $\mathbf{n}$ as axis,

$$\mathbf{Rn} = R\cos(\mathbf{R}, \mathbf{n}).$$

By what has been said we see that if $\mathbf{A}$, $\mathbf{B}$ be rigidly linked together and thus moved about in space in any arbitrary manner whatever (spun round, etc.), the value of the product $\mathbf{AB}$ is not changed. It is thus an invariant of the pair of vectors with respect to their common rigid motion. In fact, $\mathbf{AB}$ depends only on the tensors of $\mathbf{A}$, $\mathbf{B}$ and on their *relative* direction, *i.e.* the angle $(\mathbf{A}, \mathbf{B})$.

By the fusion of $\mathbf{A}$, $\mathbf{B}$ into $\mathbf{AB}$ all directional properties of the factors are gone. The result has nothing more to do with direction in space; it is an ordinary scalar, like the tensor of each of the two vectorial factors. Thus, if $\mathbf{C}$ be a third vector,

$$(\mathbf{AB})\mathbf{C} \quad \text{or} \quad \mathbf{C}(\mathbf{AB})$$

will simply mean the vector $\mathbf{C}$ magnified (stretched) $\mathbf{AB}$ times, assuming, that is, that $\mathbf{AB}$ is a dimensionless or pure number; if $\mathbf{AB}$ is an area and $\mathbf{C}$, say, a displacement, then $(\mathbf{AB})\,C$, the tensor of $(\mathbf{AB})\mathbf{C}$, is a volume, of course, and so on. If $\mathbf{D}$ is a fourth vector,

$$(\mathbf{AB})(\mathbf{CD})$$

will again be a scalar, and so on. The brackets are here used as *separators*. They are, of course, indispensable in such and similar cases. For, to take only three factors, $\mathbf{ABC}$ would, in general, be ambiguous, since

$$(\mathbf{AB})\mathbf{C} \text{ is a vector along } \mathbf{C},$$

while

$$\mathbf{A}(\mathbf{BC}) \text{ is a vector along } \mathbf{A},$$

and thus entirely different from the former. Instead of brackets *dots* may conveniently be used as separators, thus

$$(\mathbf{AB})\mathbf{C} = \mathbf{AB} \,.\, \mathbf{C},$$
$$(\mathbf{AB})(\mathbf{CD}) = \mathbf{AB} \,.\, \mathbf{CD},$$

and so forth. The reader will soon find that this need of precaution gives rise to no serious inconvenience.

The scalar product $\mathbf{AB}$ is commutative owing to the symmetry of its very definition with respect to $\mathbf{A}$, $\mathbf{B}$. In this it resembles the ordinary product. But, what is most important, it has also the *distributive* property, viz. for any $\mathbf{A}$, $\mathbf{B}$, $\mathbf{C}$,

$$\mathbf{A}(\mathbf{B}+\mathbf{C}) = \mathbf{AB} + \mathbf{AC}. \tag{15}$$

For, by the definition, $\mathbf{A}(\mathbf{B}+\mathbf{C})$ or $(\mathbf{B}+\mathbf{C})\mathbf{A}$ is the projection of the vector $\mathbf{B}+\mathbf{C}$ upon $\mathbf{A}$ multiplied by A. But the projection of the sum of two (or more) vectors upon any axis is equal to the

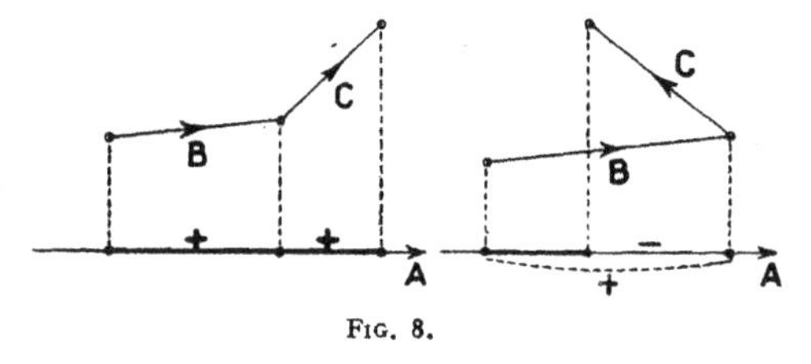

FIG. 8.

algebraic sum of the projections (Fig. 8), whence the proof of the distributive law (15).

Similarly,

$$\mathbf{A}(\mathbf{B}+\mathbf{C}+\mathbf{D}+\mathbf{E}+\ldots) = \mathbf{AB}+\mathbf{AC}+\mathbf{AD}+\mathbf{AE}+\ldots,$$

and also

$$\begin{aligned}(\mathbf{A}+\mathbf{B})(\mathbf{C}+\mathbf{D}) &= (\mathbf{A}+\mathbf{B})\mathbf{C} + (\mathbf{A}+\mathbf{B})\mathbf{D} \\ &= \mathbf{C}(\mathbf{A}+\mathbf{B}) + \mathbf{D}(\mathbf{A}+\mathbf{B}) = \mathbf{AC}+\mathbf{BC}+\mathbf{AD}+\mathbf{BD}.\end{aligned}$$

And since $\mathbf{B}-\mathbf{C}$ is the same thing as $\mathbf{B}+(-\mathbf{C})$, we have also

$$\mathbf{A}(\mathbf{B}-\mathbf{C})=\mathbf{AB}-\mathbf{AC}.$$

In fine, *the scalar multiplication of vectors is commutative as well as distributive*, and *any two vector polynomials are multiplied out precisely as in ordinary algebra.* This makes the scalar multiplication of vectors a powerful operation.

As examples we may quote

$$(\mathbf{A}+\mathbf{B})(\mathbf{A}-\mathbf{B})=\mathbf{A}^2-\mathbf{B}^2=A^2-B^2,$$

meaning that the product of the lengths of the diagonals of a parallelogram multiplied by the cosine of their included angle is

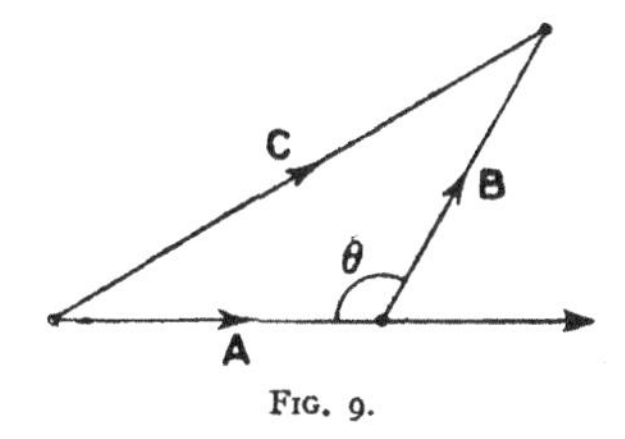

FIG. 9.

equal to the difference of the squares constructed upon the sides of the parallelogram ; again,

$$(\mathbf{A}+\mathbf{B})^2=A^2+B^2+2\mathbf{AB},$$

or (Fig. 9), remembering that $\mathbf{AB}=AB\cos(\pi-\theta)=-AB\cos\theta$,

$$C^2=A^2+B^2-2AB\cos\theta,$$

the well-known trigonometrical relation. In particular, if $\mathbf{A}\perp\mathbf{B}$,

$$(\mathbf{A}+\mathbf{B})^2=A^2+B^2,$$

the theorem of Pythagoras. As a third example, let us quote the scalar product of two coinitial unit vectors, written as in (7*a*),

$$\mathbf{r}_1=[\mathbf{i}\cos\phi_1+\mathbf{j}\sin\phi_1]\sin\theta_1+\mathbf{k}\cos\theta_1,$$

$$\mathbf{r}_2=[\mathbf{i}\cos\phi_2+\mathbf{j}\sin\phi_2]\sin\theta_2+\mathbf{k}\cos\theta_2,$$

and representing (by their end-points) two places on the Earth * whose geographic colatitudes and longitudes are θ_1, ϕ_1 and θ_2, ϕ_2. If s be their geodesic or shortest distance, *i.e.* the angle $(\mathbf{r}_1, \mathbf{r}_2)$, we have $\cos s=\mathbf{r}_1\mathbf{r}_2$. Now $\mathbf{i}^2=1$, etc., and $\mathbf{ij}=\mathbf{jk}=\mathbf{ki}=0$. Thus,

* Assumed to be ideally spherical, of radius taken for unit length.

multiplying out the two trinomials we have, for the required distance s,

$$\cos s = \cos \theta_1 \,.\, \cos \theta_2 + \sin \theta_1 \,.\, \sin \theta_2 (\cos \phi_1 \,.\, \cos \phi_2 + \sin \phi_1 \,.\, \sin \phi_2).$$

Again, calling for the moment **a**, **b** two equatorial unit vectors having the longitudes of the two places (cf. Fig. 6), viz.

$$\mathbf{a} = \mathbf{i} \,.\, \cos \phi_1 + \mathbf{j} \,.\, \sin \phi_1, \quad b = \mathbf{i} \cos \phi_2 + \mathbf{j} \sin \phi_2,$$

we have

$$\mathbf{ab} = \cos (\phi_1 - \phi_2) = \cos \phi_1 \cos \phi_2 + \sin \phi_1 \sin \phi_2,$$

the well-known formula of plane trigonometry, so that the geodesic distance of the two places, (θ_1, ϕ_1) and (θ_2, ϕ_2), becomes

$$\cos s = \cos \theta_1 \cos \theta_2 + \sin \theta_1 \sin \theta_2 \cos (\phi_1 - \phi_2), \tag{7b}$$

an important formula for navigators, which is at the same time the fundamental "cosine formula" of spherical trigonometry. In fact, N being the pole $(\theta = 0)$, formula (7*b*) concerns the spherical triangle 1*N*2 (Fig. 10), whose sides are s, θ_1, θ_2, and whose angle

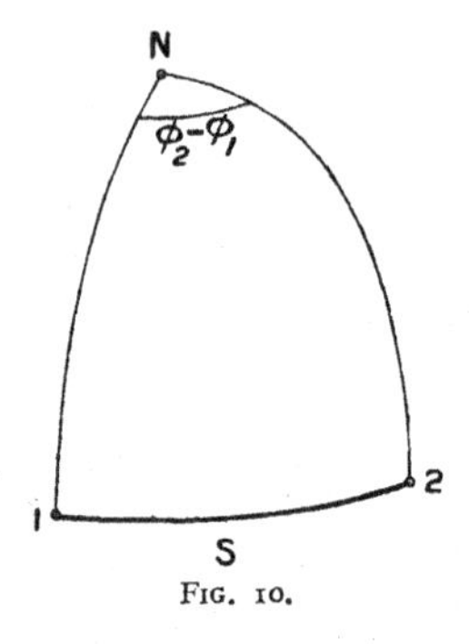

FIG. 10.

included between the latter two is $\phi_2 - \phi_1$. Notice that this is valid for any spherical triangle; for one of its corners can always be considered as our pole, $\theta = 0$.

The reader will not be astonished to see the comparatively complicated theorems of euclidean geometry thus to follow without the least trouble from squaring the sum of vectors or from multiplying scalarly two unit vectors. For essentially all euclidean relations have been condensed into the above vectorial definitions and rules of operations (addition and scalar multiplication). Still, as such a condensed system, the vector algebra is exceedingly useful. The reader will find for himself that the vector equality and the vector addition alone, as explained in Sections 1 to 4, even without the help of the scalar product, are sufficient to demonstrate formally a large number of euclidean theorems, such, for instance, as the mutual bisection of the diagonals of a parallelogram, the common cross of the three medians of a triangle, and so on.

The scope and purpose of this booklet do not permit us to enter into all these attractive details. The willing reader will, however, find no difficulty in treating them as exercises which he will soon find to be easy as well as interesting and useful, when skill in handling the vector method is aimed at.

6. The vector product of vectors. Two non-collinear vectors, **A** and **B**, can always be said to define a plane **A**, **B**, by making them coinitial, for instance, as in Fig. 11. We already know that one

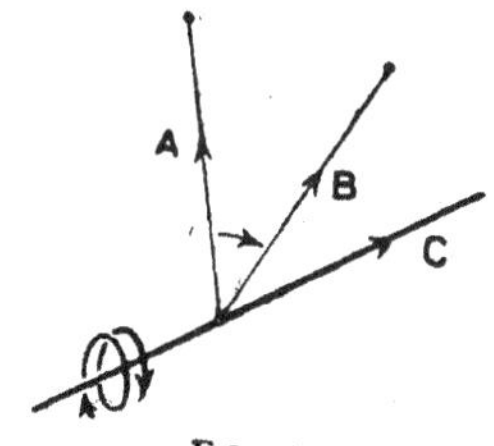

FIG. 11.

of the previous operations, **AB**, deprives them of all their properly vectorial characteristics, and the other, **A** + **B**, or more generally $x\mathbf{A}+y\mathbf{B}$, gives us only vectors which are again in the plane **A**, **B**.

The operation to be now introduced is in this respect particularly interesting, since it yields a vector outside the plane of the operands **A**, **B**.

Definition. We call *vector product of* **A** *into* **B** and denote by **VAB** a third vector **C** *normal* to **A**, **B** and drawn so that for an observer glancing along **C** the rotation turning **A** into **B**, through an angle *smaller than* 180°, is *clockwise.* This fixes the direction and the sense of the vector product **C** = **VAB**, and its tensor is defined as equal to *the area* of the parallelogram constructed upon **A**, **B** as sides, *i.e.*

$$C=|\mathbf{VAB}|=AB\,|\sin(\mathbf{A},\mathbf{B})|. \qquad (16)$$

From this definition we see, first of all, that the vector product is *not* commutative, inasmuch as we have

$$\mathbf{VBA}=-\mathbf{VAB}. \qquad (17)$$

Again, if $\mathbf{A}, \mathbf{B}$ are parallel to one another, *i.e.* collinear, we have

$$\mathrm{V}\mathbf{AB} = 0.$$

And if $\mathbf{A} \perp \mathbf{B}$, then $\sin(\mathbf{A}, \mathbf{B}) = \pm 1$, and

$$C = |\mathrm{V}\mathbf{AB}| = AB,$$

while $\mathbf{A}$, $\mathbf{B}$ and $\mathbf{C}$ form a *right-handed* normal system of three vectors. If $\mathbf{A}$ points upward and $\mathbf{B}$ towards the right, then $\mathbf{C} = \mathrm{V}\mathbf{AB}$ points forward.

If we know of two vectors $\mathbf{A}, \mathbf{B}$ that their vector product vanishes, then we can conclude only that they are parallel (collinear), *i.e.* that

$$\mathbf{A} = m\mathbf{B},$$

where m is some undetermined scalar number, but by no means that one of the vectors vanishes (unless we know beforehand that they cannot be parallel). This is, *mutatis mutandis*, analogous to what has been said in Section 5 with regard to the scalar product.

From (16) we see at once that the vector product of m times $\mathbf{A}$ into n times $\mathbf{B}$ is equal to

$$mn\mathrm{V}\mathbf{AB}.$$

Thus, for instance,

$$\mathrm{V}\mathbf{AB} = AB\mathrm{V}\mathbf{ab}, \tag{18}$$

where $\mathbf{a}, \mathbf{b}$ are the units of $\mathbf{A}, \mathbf{B}$. Similarly $\mathbf{AB} = AB\mathbf{ab}$.

For a right-handed system of normal unit vectors, as the previous $\mathbf{i}, \mathbf{j}, \mathbf{k}$, we have

$$\mathrm{V}\mathbf{ij} = \mathbf{k}, \quad \mathrm{V}\mathbf{jk} = \mathbf{i}, \quad \mathrm{V}\mathbf{ki} = \mathbf{j}, \tag{a}$$

three relations derivable from one another by cyclic permutations of $\mathbf{i}, \mathbf{j}, \mathbf{k}$. At the same time we have, of course, as for every vector,

$$\mathrm{V}\mathbf{ii} = \mathrm{V}\mathbf{jj} = \mathrm{V}\mathbf{kk} = 0.$$

Contrast these relations with the previous ones, $\mathbf{i}^2 = \mathbf{j}^2 = \mathbf{k}^2 = 1$ and $\mathbf{ij} = \mathbf{jk} = \mathbf{ki} = 0$. The latter follow also from (a); for, by the second of (a), for instance, $\mathbf{i} = \mathrm{V}\mathbf{jk}$ is *normal* to $\mathbf{j}$, and therefore $\mathbf{ji} = \mathbf{j}\mathrm{V}\mathbf{jk} = 0$. It is scarcely necessary to explain that $\mathbf{j}\mathrm{V}\mathbf{jk}$ means the *scalar product* of the vectors $\mathbf{j}$ and $\mathrm{V}\mathbf{jk}$.

More generally we have, for any two vectors $\mathbf{A}, \mathbf{B}$, by the very definition of $\mathrm{V}\mathbf{AB}$,

$$\mathbf{A}\mathrm{V}\mathbf{AB} = \mathbf{B}\mathrm{V}\mathbf{AB} = 0.$$

Let now $\mathbf{A}, \mathbf{B}, \mathbf{C}$ be any three vectors whatever, generally non coplanar with one another. Then the scalarly-vectorial product,

$$\mathbf{A}\mathrm{V}\mathbf{BC},$$

which is itself a scalar, has a very simple geometrical meaning. In fact, let **A**, **B**, **C** (in the order as they are written) form a right-handed system, *i.e.* such that a person glancing along **C** sees the

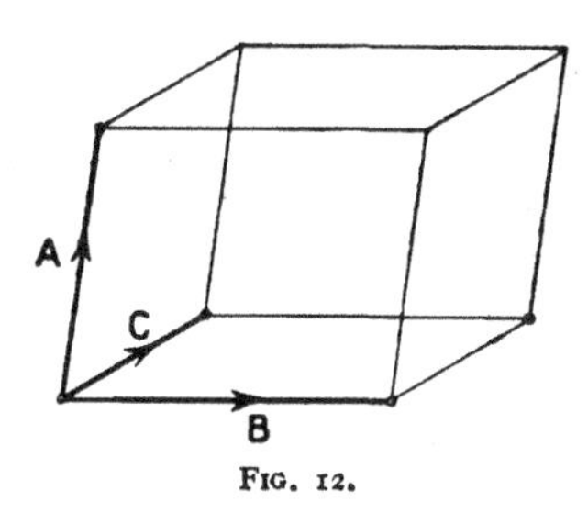

FIG. 12.

rotation from **A** to **B** (through less than π) clockwise. Construct upon **A**, **B**, **C** as edges a parallelepipedon (Fig. 12). Then **VBC** will be perpendicular to the base **B**,**C**, and its tensor will be equal to the area of this base; in symbols,

$$\mathbf{VBC} = (\text{area of base})\,\mathbf{n},$$

where **n** is a unit vector perpendicular to the base. Therefore,

$$\mathbf{AVBC} = (\text{area of base})\,\mathbf{An},$$

and **An** being the height of the parallelepipedon, we see that

$$\mathbf{AVBC} = \text{volume of parallelepipedon } \mathbf{A},\ \mathbf{B},\ \mathbf{C},$$

provided that **A**, **B**, **C** is a right-handed arrangement of the edges. (If it were a left-handed arrangement, then **AVBC** would be equal to *minus* the volume.) Now, the same volume can be expressed by taking **C**, **A** or **A**, **B** as base. Thus we obtain the important property

$$\mathbf{AVBC} = \mathbf{BVCA} = \mathbf{CVAB}, \tag{19}$$

or in words: the cyclic permutation of the three factors of **AVBC** does not influence the value of the product.* Inverting the cyclic order is equivalent to changing its sign. For **VCB** is minus **VBC**. The particular property $(x\mathbf{A} + y\mathbf{B})\mathbf{VAB} = 0$ can now be interpreted geometrically by saying that the volume of a parallelepipedon

* The validity of formula (19) is by no means based upon this volume-proof (or rather illustration), which is given here only because it best appeals to simple intuitions. In fact, (19) can be proved algebraically, without any appeal to the concept of 'volume.'

vanishes when its three edges become coplanar, that is to say, when all its faces collapse into one plane.

If of any three vectors $\mathbf{A}$, $\mathbf{B}$, $\mathbf{C}$ we know that

$$\mathbf{A}\mathrm{V}\mathbf{BC} = 0,$$

then the only thing we can conclude is that $\mathbf{A}$, $\mathbf{B}$, $\mathbf{C}$ are *coplanar*, but by no means that one of these vectors vanishes. Conversely, if $\mathbf{A}$, $\mathbf{B}$, $\mathbf{C}$ are coplanar, we have $\mathbf{A}\mathrm{V}\mathbf{BC} = 0$. The theorem expressed by (19) is of great utility in many applications, and it deserves, therefore, to be well kept in mind.

As in the case of the scalar product, one of the most important properties of *the vector product* is its *distributivity*, *i.e.* for any three vectors $\mathbf{A}$, $\mathbf{B}$, $\mathbf{C}$,

$$\mathrm{V}\mathbf{A}(\mathbf{B}+\mathbf{C}) = \mathrm{V}\mathbf{AB} + \mathrm{V}\mathbf{AC}. \tag{20}$$

This capital property can be proved in a variety of ways. First of all, by an immediate geometrical construction of both the right- and the left-hand member of (19),—which will be left as an exercise for the reader. (It will be enough if the reader constructs it for the simplest case of coplanar $\mathbf{A}$, $\mathbf{B}$, $\mathbf{C}$.*) Another, comparatively simple proof, based upon (19), is this:

Let us write

$$\mathrm{V}\mathbf{A}(\mathbf{B}+\mathbf{C}) - \mathrm{V}\mathbf{AB} - \mathrm{V}\mathbf{AC} = \mathbf{X}.$$

Then our problem is reduced to proving that $\mathbf{X}$ vanishes. Now, all the three addends being perpendicular to $\mathbf{A}$, so is their sum $\mathbf{X}$, *i.e.*

$$\mathbf{XA} = 0.$$

Again,

$$\begin{aligned} \mathbf{XB} &= \mathbf{B}\mathrm{V}\mathbf{A}(\mathbf{B}+\mathbf{C}) - \mathbf{B}\mathrm{V}\mathbf{AC} \\ &= (\mathbf{B}+\mathbf{C})\mathrm{V}\mathbf{BA} - \mathbf{C}\mathrm{V}\mathbf{BA}, \text{ by (19)}, \\ &= \mathbf{B}\mathrm{V}\mathbf{BA} + \mathbf{C}\mathrm{V}\mathbf{BA} - \mathbf{C}\mathrm{V}\mathbf{BA} = 0, \end{aligned}$$

and similarly $\mathbf{XC} = 0$. Thus, the vector $\mathbf{X}$ either vanishes or is normal to each of the three vectors $\mathbf{A}$, $\mathbf{B}$, $\mathbf{C}$. Now, if these are not coplanar, the latter case is excluded, so that $\mathbf{X} = 0$. Thus, for non-coplanar $\mathbf{A}$, $\mathbf{B}$, $\mathbf{C}$ the distributive property (20) is already proved. And if $\mathbf{A}$, $\mathbf{B}$, $\mathbf{C}$ happen to be coplanar, add to $\mathbf{C}$, for instance, a fourth vector $\mathbf{D}$ inclined to the plane of $\mathbf{A}$, $\mathbf{B}$, $\mathbf{C}$. Then the new vectors $\mathbf{A}$, $\mathbf{B}$, $\mathbf{C}+\mathbf{D}$ will not be coplanar, and

$$\mathrm{V}\mathbf{A}(\mathbf{C}+\mathbf{D}) + \mathrm{V}\mathbf{B}(\mathbf{C}+\mathbf{D}) = \mathrm{V}(\mathbf{A}+\mathbf{B})(\mathbf{C}+\mathbf{D}),$$

* For the case of non-coplanar $\mathbf{A}$, $\mathbf{B}$, $\mathbf{C}$ is more easily dealt with by the following analytical method.

and since **D** can always be so chosen as to make the three relevant vectors in each of these products non-coplanar, they may be expanded, giving

$$\mathrm{V}\mathbf{AC}+\mathrm{V}\mathbf{AD}+\mathrm{V}\mathbf{BC}+\mathrm{V}\mathbf{BD}=\mathrm{V}(\mathbf{A}+\mathbf{B})\mathbf{C}+\mathrm{V}(\mathbf{A}+\mathbf{B})\mathbf{D}\,;$$

but, by the above,

$$\mathrm{V}\mathbf{AD}+\mathrm{V}\mathbf{BD}=\mathrm{V}(\mathbf{A}+\mathbf{B})\mathbf{D},$$

whence

$$\mathrm{V}\mathbf{AC}+\mathrm{V}\mathbf{BC}=\mathrm{V}(\mathbf{A}+\mathbf{B})\mathbf{C},$$

or, changing the sign of both sides,

$$\mathrm{V}\mathbf{CA}+\mathrm{V}\mathbf{CB}=\mathrm{V}\mathbf{C}(\mathbf{A}+\mathbf{B}).$$

Thus the distributive property of vector multiplication is proved for any **A**, **B**, **C**, coplanar or not.

The product of two binomials (or polynomials) does not call for lengthy explanations. Thus,

$$\begin{aligned}\mathrm{V}(\mathbf{A}+\mathbf{B})(\mathbf{C}+\mathbf{D})&=\mathrm{V}(\mathbf{A}+\mathbf{B})\mathbf{C}+\mathrm{V}(\mathbf{A}+\mathbf{B})\mathbf{D}\\&=-\mathrm{V}\mathbf{C}(\mathbf{A}+\mathbf{B})-\mathrm{V}\mathbf{D}(\mathbf{A}+\mathbf{B})=\mathrm{V}\mathbf{AC}+\mathrm{V}\mathbf{BC}+\mathrm{V}\mathbf{AD}+\mathrm{V}\mathbf{BD}.\end{aligned}$$

The vector multiplication of any two vector polynomials is thus seen to obey the same rules as ordinary algebraic multiplication, the only difference being that vector products are *not* commutative. A reversal of the order of the two factors changes only the sign of their product, which is easily remembered.

7. Expansion of vector formulae. Basing ourselves upon the distributive property just proved, we can at once expand the vector product of any two vectors into its cartesian or any other form. Thus, if

$$\mathbf{A}=A_1\mathbf{i}+A_2\mathbf{j}+A_3\mathbf{k}, \quad \text{and} \quad \mathbf{B}=B_1\mathbf{i}+B_2\mathbf{j}+B_3\mathbf{k},$$

we have, remembering that $\mathrm{V}\mathbf{ii}=0$, $\mathrm{V}\mathbf{jk}=-\mathrm{V}\mathbf{kj}=\mathbf{i}$, etc.,

$$\mathrm{V}\mathbf{AB}=\mathbf{i}(A_2B_3-A_3B_2)+\mathbf{j}(A_3B_1-A_1B_3)+\mathbf{k}(A_1B_2-A_2B_1), \quad (21)$$

exhibiting $A_2B_3-A_3B_2$, etc. (by cyclic permutation), as the three rectangular components of the vector product. Since $|\mathrm{V}\mathbf{AB}|$ is the area of the parallelogram constructed upon **A**, **B** as sides, we see at the same time that $A_2B_3-A_3B_2$, etc., are the areas of the projections of this parallelogram upon the planes **j**, **k**; **k**, **i**; **i**, **j**,— a well-known result which, however, is more easily seen on the vector method. The last formula, (21), is easily memorized in its determinantal form, which is

$$\mathbf{VAB} = \begin{vmatrix} \mathbf{i} & \mathbf{j} & \mathbf{k} \\ A_1 & A_2 & A_3 \\ B_1 & B_2 & B_3 \end{vmatrix}. \tag{21a}$$

In exactly the same way the reader will show himself that the cartesian expansion of **AVBC**, the triple product representing the volume of the parallelepipedon **A**, **B**, **C**, is

$$\mathbf{AVBC} = \begin{vmatrix} A_1 & A_2 & A_3 \\ B_1 & B_2 & B_3 \\ C_1 & C_2 & C_3 \end{vmatrix}. \tag{22}$$

This, in fact, is the most familiar expression for the volume of the parallelepipedon constructed upon **A**, **B**, **C** as edges. Formula (22) gives also an immediate verification of the property

$$\mathbf{AVBC} = \mathbf{BVCA}, \text{ etc.,}$$

as in (19). For

$$\begin{vmatrix} A_1 & A_2 & A_3 \\ B_1 & B_2 & B_3 \\ C_1 & C_2 & C_3 \end{vmatrix} = \begin{vmatrix} B_1 & B_2 & B_3 \\ C_1 & C_2 & C_3 \\ A_1 & A_2 & A_3 \end{vmatrix},$$

and so on.

For the scalar product we have immediately, remembering that $\mathbf{i}^2 = 1$, $\mathbf{ij} = 0$, etc.,

$$\mathbf{AB} = A_1B_1 + A_2B_2 + A_3B_3. \tag{23}$$

As particular cases of (21) and (23) note the results for two unit vectors **a**, **b** which include the angle ϖ,

$$\sin^2\varpi = (a_2b_3 - a_3b_2)^2 + (a_3b_1 - a_1b_3)^2 + (a_1b_2 - a_2b_1)^2,$$

$$\cos\varpi = a_1b_1 + a_2b_2 + a_3b_3,$$

a_1, b_1, etc., being now the direction-cosines of **a**, **b** relatively to **i**, **j**, **k** as axes. For such is the meaning of the components of *unit* vectors.

In order to give at least one illustration of the utility of **AVBC**, let us consider three coinitial unit vectors whose end-points may be conceived as the vertices 1, 2, 3 of a spherical triangle drawn on a unit sphere. Let us use the colatitude and the longitude as in (7*a*). Without any loss to generality we may put the pole

$(\theta=0)$ into the vertex 1 and take the first meridian along the side 12; thus, α_1 being the angle at 1, and s_2, s_3 the sides of the spherical triangle opposite 2 and 3,

$$\mathbf{r}_1=\mathbf{i},$$
$$\mathbf{r}_2=\mathbf{i}\cos s_3+\mathbf{j}\sin s_3,$$
$$\mathbf{r}_3=\mathbf{i}\cos s_2+\sin s_2\,.\,[\mathbf{j}\cos\alpha_1+\mathbf{k}\sin\alpha_1].$$

This gives for the scalarly-vectorial product, by (22), since the first vector has no second and no third component, and the second vector no third component,

$$\mathbf{r}_1\mathrm{V}\mathbf{r}_2\mathbf{r}_3=\sin s_2\sin s_3\sin\alpha_1,$$

which, by the cyclical property (19), is also equal to $\mathbf{r}_2\mathrm{V}\mathbf{r}_3\mathbf{r}_1$ and to $\mathbf{r}_3\mathrm{V}\mathbf{r}_1\mathbf{r}_2$, and these products are obviously equal to

$$\sin s_3\sin s_1\sin\alpha_2$$

and to

$$\sin s_1\sin s_2\sin\alpha_3,$$

where α_2, α_3 are the remaining two angles of the spherical triangle.

Thus,

$$\frac{\sin\alpha_1}{\sin s_1}=\frac{\sin\alpha_2}{\sin s_2}=\frac{\sin\alpha_3}{\sin s_3}, \tag{19a}$$

the fundamental "sine formula" of spherical trigonometry, following on the vector method as easily as the "cosine formula" given before. It is interesting to note that the "sine formula" is, in this circle of ideas, but the statement of the triple expressibility of the volume of the parallelepipedon $\mathbf{r}_1$, $\mathbf{r}_2$, $\mathbf{r}_3$, viz. as $\mathbf{r}_1\mathrm{V}\mathbf{r}_2\mathbf{r}_3$, or $\mathbf{r}_2\mathrm{V}\mathbf{r}_3\mathbf{r}_1$ or $\mathbf{r}_3\mathrm{V}\mathbf{r}_1\mathbf{r}_2$. Other examples are left to the care of the reader.

8. Iteration of vectorial multiplication. There is but one more important formula to be noted in connection with the vector product of vectors, viz. a formula giving a convenient vector expansion of the result of repeated or iterated vector multiplication,

$$\mathrm{V}\mathbf{A}(\mathrm{V}\mathbf{BC})\quad\text{or simply}\quad\mathrm{V}\mathbf{A}\mathrm{V}\mathbf{BC},$$

which reads: having obtained the vector product of **B**, **C**, multiply it, again vectorially, by **A**. This ternary product, which occurs very often, is, of course, again a vector, to wit, perpendicular to **A** and to $\mathrm{V}\mathbf{BC}$; but the latter being itself perpendicular to **B**, **C**, our new vector $\mathbf{VAVBC}$ is coplanar with $\mathbf{B}$, $\mathbf{C}$, so that we know beforehand that the result will be of the form *

$$\mathbf{VAVBC} = \beta\mathbf{B} + \gamma\mathbf{C},$$

where β, γ are some scalars. Since the ternary product is perpendicular to $\mathbf{A}$, we have $\beta(\mathbf{AB}) + \gamma(\mathbf{AC}) = 0$, so that

$$\mathbf{VAVBC} = \lambda\{\mathbf{B}(\mathbf{CA}) - \mathbf{C}(\mathbf{AB})\},$$

where λ is a scalar. It remains to determine its numerical value. This can be done, for instance, in the following manner. First of all, $\mathbf{A}$ can always be assumed to be coplanar with $\mathbf{B}$, $\mathbf{C}$, since its part normal to $\mathbf{B}$, $\mathbf{C}$ contributes nothing. Next, dividing both sides by ABC, the equation becomes

$$\mathbf{VaVbc} = \lambda\{\mathbf{b}(\mathbf{ca}) - \mathbf{c}(\mathbf{ab})\},$$

where $\mathbf{a}$, $\mathbf{b}$, $\mathbf{c}$ are the units of $\mathbf{A}$, $\mathbf{B}$, $\mathbf{C}$. Now, multiply both sides scalarly by $\mathbf{b}$, and notice that, by (19),

$$\mathbf{bVaVbc} = (\mathbf{Vbc})(\mathbf{Vba}) = \sin(\mathbf{b}, \mathbf{c}) \,.\, \sin(\mathbf{b}, \mathbf{a}).$$

Thus,

$$\sin(\mathbf{b}, \mathbf{c}) \,.\, \sin(\mathbf{b}, \mathbf{a}) = \lambda[\cos(\mathbf{c}, \mathbf{a}) - \cos(\mathbf{a}, \mathbf{b}) \,.\, \cos(\mathbf{b}, \mathbf{c})] ;$$

but, the three vectors being coplanar, we have

$$\cos(\mathbf{c}, \mathbf{a}) = \cos(\mathbf{b}, \mathbf{a}) \,.\, \cos(\mathbf{b}, \mathbf{c}) + \sin(\mathbf{b}, \mathbf{a}) \,.\, \sin(\mathbf{b}, \mathbf{c}),$$

so that $\lambda = 1$.

The required formula is, therefore,

$$\mathbf{VAVBC} = \mathbf{B}(\mathbf{CA}) - \mathbf{C}(\mathbf{AB}). \tag{24}$$

As an exercise, the reader may verify it by an iterated application of the cartesian expansion (21) or (21*a*). Having once obtained this important formula, there will be no difficulty in dealing with quaternary vector products, as $\mathbf{VDVAVBC}$, which becomes $(\mathbf{CA})\mathbf{VDB} - (\mathbf{AB})\mathbf{VDC}$, etc. But such products will hardly occur in practice.

A notable property of the above ternary product and of its two cyclical permutations is that

$$\mathbf{VAVBC} + \mathbf{VBVCA} + \mathbf{VCVAB} = 0, \tag{24a}$$

identically. For the six right-hand terms of (24) and of the two similar equations destroy themselves in pairs.

* The trivial case of $\mathbf{B}$, $\mathbf{C}$ collinear can be discarded ; for then $\mathbf{VAVBC} = 0$.

A particular case of (24) which often occurs is that in which $\mathbf{C}$ is equal to $\mathbf{A}$ and is an unit vector $\mathbf{u}$, say. Then we have

$$\mathrm{V}\mathbf{u}\mathrm{V}\mathbf{B}\mathbf{u} = \mathbf{B} - (\mathbf{B}\mathbf{u})\mathbf{u}, \tag{24b}$$

whence we see also that $\mathrm{V}\mathbf{u}\mathrm{V}\mathbf{B}\mathbf{u}$ is the part of the vector $\mathbf{B}$ *normal* to $\mathbf{u}$, in both size and direction. For $(\mathbf{B}\mathbf{u})\mathbf{u}$ is the part of $\mathbf{B}$ along $\mathbf{u}$.

To close this section, and at the same time the essential part of the whole Vector Algebra, but a few more remarks which will be useful in connection with problems often occurring in practice.

Let $\mathbf{X}$ be an unknown, and $\mathbf{A}$, $\mathbf{u}$ two given vectors, the latter an unit vector. If we know of $\mathbf{X}$ only that

$$\mathrm{V}\mathbf{X}\mathbf{u} = \mathbf{A}, \tag{a}$$

we cannot fully determine $\mathbf{X}$. For to a solution of this equation we can add any vector $m\mathbf{u}$ (since $\mathrm{V}\mathbf{u}\mathbf{u} = 0$), and $\mathbf{X} + m\mathbf{u}$ will again be a solution of this equation. In order to determine $\mathbf{X}$ uniquely we must have one more (scalar) datum. Let this be

$$\mathbf{X}\mathbf{u} = \sigma, \tag{b}$$

where σ is a given scalar. Then $\mathbf{X}$ is completely determined. In order to find its value explicitly in terms of the given $\mathbf{A}$, $\mathbf{u}$, σ, multiply the equation (a) vectorially by $\mathbf{u}$; then, in virtue of (24),

$$\mathbf{X} - (\mathbf{X}\mathbf{u})\mathbf{u} = \mathrm{V}\mathbf{u}\mathbf{A},$$

and by (b),

$$\mathbf{X} = \sigma\mathbf{u} + \mathrm{V}\mathbf{u}\mathbf{A}, \tag{c}$$

which is the required solution. This simple rule, (c), for solving the equations (a) and (b), will often be found helpful.

9. The Linear Vector Operator. Let $\mathbf{R}$ be a variable vector, that is to say, one that can assume in turn all possible sizes (tensors) and directions. Of each of these determined vectors we can speak as of the special value of the variable $\mathbf{R}$. To have a good picture of such an abstract concept, imagine $\mathbf{R}$ as a straight, extensible and contractile string fixed at one of its ends at a permanent point O; then its free end-point P occupying in succession all possible points of space, OP will represent the various values of $\mathbf{R}$. The vector $\mathbf{R}$ can, in such a connection, be advantageously called the *position vector* of the point or, if one prefers, of the particle P. Now imagine that there is another particle P', and let its position vector, with the same origin O, be called $\mathbf{R}'$. Let there be some mechanism, or else our own imagination, which to every chosen position of P makes correspond a certain position of P'. This we may express by saying that to every value of $\mathbf{R}$ corresponds a certain value of $\mathbf{R}'$, by writing

$$\mathbf{R}' = \varpi\mathbf{R}, \tag{25}$$

and by calling $\mathbf{R}'$ a *vector function* of the variable vector $\mathbf{R}$. If, as we assume, to every $\mathbf{R}$ corresponds but one $\mathbf{R}'$, determined in size and direction, we will say that $\mathbf{R}'$ is a *monovalent* function of $\mathbf{R}$, and we will call ϖ a monovalent vector *operator*, the symbol of some operations to be performed on $\mathbf{R}$ in order to obtain $\mathbf{R}'$. We can think of such operations in the algebraical, as well as in the physical sense of the word, as turning round the representative string, stretching or contracting it according to some more or less complicated prescription. It is needless to explain that an equation such as (25) is equivalent to three scalar equations: each of the components of $\mathbf{R}'$ equal to some function of, in general, all the three components of $\mathbf{R}$.

Suppose now that $\mathbf{R}$ is represented as the sum $\mathbf{A}+\mathbf{B}$ of some two vectors. In general the operations embodied in ϖ may be such that $\varpi(\mathbf{A}+\mathbf{B})$ is not the same thing as $\varpi\mathbf{A}+\varpi\mathbf{B}$. A good example of such an operator is that which converts an incident luminous ray into the refracted ray (cf. *Simplified Method*, quoted in Preface). But the operations represented by ϖ may also, in particular, be such that

$$\varpi(\mathbf{A}+\mathbf{B}) = \varpi\mathbf{A}+\varpi\mathbf{B},$$

whatever the vectors $\mathbf{A}$ and $\mathbf{B}$. If such be the case we call ϖ a *distributive* operator. An example of this kind is afforded by the "reflector," *i.e.* that operator which converts the incident ray into the reflected one. The simplest example of a distributive vector operator is, however, a scalar number σ used as a factor; for we have, of course,

$$\sigma(\mathbf{A}+\mathbf{B}) = \sigma\mathbf{A}+\sigma\mathbf{B}.$$

This operator is a pure stretcher or (if $|\sigma| < 1$) a contractor, and, if $\sigma < 0$, an invertor at the same time.

Leaving these examples, let us turn to the general distributive operator, of which we will only assume that it is a *continuous* operator, *i.e.* that $\varpi\mathbf{R}$ is a continuous vector function of $\mathbf{R}$. Such distributive operators have very far-reaching applications in many branches of geometry and physics. They are known better under the name of *linear vector operators*, and were first introduced by the great Hamilton. In fact, it can easily be shown that the continuous and distributive operator ϖ when applied to a vector $\mathbf{R}$ yields another vector $\mathbf{R}'$, whose components are each a *linear function* of the components of $\mathbf{R}$, whatever the triad of axes employed for the decomposition. For, if n be any integer positive scalar number, we have, in virtue of the assumed distributive property,

$$\varpi(n\mathbf{A}) = n\varpi(\mathbf{A}),$$

and this property can easily be extended to negative and fractional values of n, and ultimately, by the often repeated limit-reasoning, to any real value of n. This being granted, let $\mathbf{a}$, $\mathbf{b}$, $\mathbf{c}$ be any *non-coplanar* unit vectors; then, whatever $\mathbf{R}$, we can represent it by

$$\mathbf{R} = x\mathbf{a} + y\mathbf{b} + z\mathbf{c},$$

where x, y, z are some scalars, the components of $\mathbf{R}$. Thus, the corresponding $\mathbf{R}'$ will be

$$\mathbf{R}' = \varpi(x\mathbf{a} + y\mathbf{b} + z\mathbf{c}) = x\varpi\mathbf{a} + y\varpi\mathbf{b} + z\varpi\mathbf{c}, \tag{26}$$

and, therefore, the components of $\mathbf{R}'$, *i.e.* $x' = \mathbf{a}\mathbf{R}'$, etc.,

$$\left.\begin{aligned} x' &= (\mathbf{a}\varpi\mathbf{a})x + (\mathbf{a}\varpi\mathbf{b})y + (\mathbf{a}\varpi\mathbf{c})z, \\ y' &= (\mathbf{b}\varpi\mathbf{a})x + \text{etc.}; \quad z' = (\mathbf{c}\varpi\mathbf{a})x + \text{etc.}, \end{aligned}\right\} \tag{26a}$$

and these are linear, homogeneous functions of x, y, z, the components of $\mathbf{R}$,—the coefficients $(\mathbf{a}\varpi\mathbf{a})$, etc., being certain scalars (scalar products of $\mathbf{a}$ and $\varpi\mathbf{a}$, $\mathbf{a}$ and $\varpi\mathbf{b}$, and so on), which depend partly on the nature of the operator ϖ, and partly on the choice of the framework $\mathbf{a}$, $\mathbf{b}$, $\mathbf{c}$.

From (26*a*) we see that the operator ϖ is fully determined if we give, for any chosen $\mathbf{a}$, $\mathbf{b}$, $\mathbf{c}$, the *nine* coefficients

$$\mathbf{a}\varpi\mathbf{a}, \quad \mathbf{a}\varpi\mathbf{b}, \quad \mathbf{a}\varpi\mathbf{c}, \quad \mathbf{b}\varpi\mathbf{a}, \text{ etc.},$$

which we will denote by

$$\varpi_{aa}, \quad \varpi_{ab}, \quad \varpi_{ac}, \quad \varpi_{ba}, \text{ etc.},$$

respectively. And it is not difficult to see that these nine scalar data, which are sufficient, are also necessary to determine completely a linear vector operator.

We can express the same thing more simply by taking the vector equation (26) instead of its components (26*a*). Let us rewrite (26), taking for **a**, **b**, **c** any triad of *normal* vectors; then x, y, z stand for **aR**, etc., so that the equation is

$$\mathbf{R}' = \varpi\mathbf{a}(\mathbf{aR}) + \varpi\mathbf{b}(\mathbf{bR}) + \varpi\mathbf{c}(\mathbf{cR}). \tag{26'}$$

Thus we can say that the operator ϖ is fully determined if we know what it yields if applied to the three conventional vectors **a**, **b**, **c**, that is to say, if we are given the three vectors

$$\varpi\mathbf{a}, \quad \varpi\mathbf{b}, \quad \varpi\mathbf{c}.$$

Each of these implies 3 scalar data, so that in all we have again 9 data, as before. What we have, a moment ago, denoted by ϖ_{aa}, ϖ_{ab}, etc., are simply the cartesian components of these three vector data, taken along the axes **a**, **b**, **c**.

Hitherto we have spoken of the most general linear vector operator. Let us now explain an important subdivision of this vast class of operators. Let **A** and **B** be any two vectors. We can take $\varpi\mathbf{B}$ and multiply it scalarly by **A**, or first form $\varpi\mathbf{A}$ and then multiply it so by **B**. In this way we should obtain the two scalars,

$$\mathbf{A}\varpi\mathbf{B} \quad \text{and} \quad \mathbf{B}\varpi\mathbf{A}.$$

Now, in general, these two numbers will be different from one another. But the operator ϖ may happen to be such that they are equal, *i.e.* that, *for any* **A**, **B**,

$$\mathbf{A}\varpi\mathbf{B} = \mathbf{B}\varpi\mathbf{A}. \tag{27}$$

If such be the case we call ϖ a *self-conjugate* operator. By this definition (27) we shall also have $\varpi_{ab} = \varpi_{ba}$, etc., so that a self-conjugate operator has but *six* mutually independent (that is, six independently prescribable) coefficients, or constituents,

$$\begin{array}{ccc} \varpi_{aa} & \varpi_{ab} & \varpi_{ac} \\ \cdot & \varpi_{bb} & \varpi_{bc} \\ \cdot & \cdot & \varpi_{cc}. \end{array}$$

This table, after the insertion of $\varpi_{ba} = \varpi_{ab}$, etc., at the vacant places, is symmetrical with respect to its diagonal; whence also the name of *symmetrical* operator, used as a synonym for the self-conjugate operator.

This being a sub-class of ϖ, the general operator, the remainder of the class of ϖ's, for which

$$\varpi_{ab} \neq \varpi_{ba}, \text{ etc.,}$$

are called non-symmetrical or *asymmetrical* operators. If ϖ be such an operator, and if ϖ' be another operator such that

$$\mathbf{A}\varpi\mathbf{B} = \mathbf{B}\varpi'\mathbf{A}, \tag{28}$$

for any pair **A**, **B** of vectors, then ϖ' is called the *conjugate* of ϖ. Obviously also ϖ is the conjugate of ϖ'. Thus, if ϖ'_{ab}, etc., be the coefficients of ϖ', we have, remembering that ϖ_{ab} stands for $\mathbf{a}\varpi\mathbf{b}$, etc.,

$$\varpi'_{ab} = \varpi_{ba}, \quad \varpi'_{bc} = \varpi_{cb}, \quad \varpi'_{ca} = \varpi_{ac}, \tag{28a}$$

while, of course, $\varpi'_{aa} = \varpi_{aa}$, etc. Thus we see also that to every operator ϖ there is one (and only one) conjugate ϖ'. In particular, if ϖ is a symmetrical operator, its conjugate is identical with it, whence "self-conjugate" as a synonym of symmetrical operators. In harmony with this, (27) is but a special case of (28), viz. for $\varpi' = \varpi$.

Let us use ϖ for any linear vector operator, and ω for *symmetrical* operators only. (In fact, without the circumflex this last letter of the Greek alphabet has some symmetry.)

Manifestly the symmetrical operator ω will be a great deal simpler than the asymmetrical ϖ. It is, therefore, very agreeable to see that any ϖ can be split into an ω and some other asymmetrical, but very simple, operator which is called an *antisymmetrical* (or skew-) operator and which we will denote by α. The latter is defined most conveniently by saying that, for any **A**, **B**,

$$\mathbf{A}\alpha\mathbf{B} = -\mathbf{B}\alpha\mathbf{A}, \quad \therefore \ \mathbf{A}\alpha\mathbf{A} = 0, \tag{29}$$

and therefore also $\alpha_{ab} = -\alpha_{ba}$, etc., and $\alpha_{aa} = 0$, etc., so that the table for such an operator becomes

$$\begin{matrix} 0 & \alpha_{ab} & \alpha_{ac} \\ -\alpha_{ab} & 0 & \alpha_{bc} \\ -\alpha_{ac} & -\alpha_{bc} & 0 \end{matrix} \tag{29a}$$

which justifies the name. The announced property can shortly be written

$$\varpi = \omega + \alpha,$$

which is a symbolic short for

$$\varpi\mathbf{R} = \omega\mathbf{R} + \alpha\mathbf{R},$$

where **R** is any vector operand. The said property is easily proved.

In fact, let ϖ' be the conjugate of the given operator ϖ. Then we have, identically,

$$\varpi = \tfrac{1}{2}(\varpi + \varpi') + \tfrac{1}{2}(\varpi - \varpi'). \tag{30}$$

But the first term represents a symmetrical operator, because, by (28),

$$\mathbf{A}(\varpi + \varpi')\mathbf{B} = \mathbf{A}\varpi\mathbf{B} + \mathbf{A}\varpi'\mathbf{B} = \mathbf{B}\varpi'\mathbf{A} + \mathbf{B}\varpi\mathbf{A} = \mathbf{B}(\varpi + \varpi')\mathbf{A},$$

which precisely is the definition (27) of a symmetric operator. And the second term is antisymmetric, for

$$\mathbf{A}(\varpi - \varpi')\mathbf{B} = \mathbf{B}\varpi'\mathbf{A} - \mathbf{B}\varpi\mathbf{A} = -\mathbf{B}(\varpi - \varpi')\mathbf{A},$$

as in (29), the definition of antisymmetric operators. This proves the statement, without the slightest need of splitting ϖ into its nine constituents ϖ_{aa}, etc.

We thus see that every linear vector operator can be written

$$\varpi = \omega + \alpha, \tag{31}$$

where its symmetrical part is $\omega = \tfrac{1}{2}(\varpi + \varpi')$ and its antisymmetrical part $\alpha = \tfrac{1}{2}(\varpi - \varpi')$.

If the reader so desires he can introduce the nine coefficients of these operators. Then

$$\omega_{ab} = \tfrac{1}{2}(\varpi_{ab} + \varpi_{ba}) = \omega_{ba},$$

proving again that ω is self-conjugate, and

$$\alpha_{aa} = 0, \text{ etc.,} \quad \alpha_{ab} = \tfrac{1}{2}(\varpi_{ab} - \varpi_{ba}) = -\alpha_{ab},$$

proving that α is antisymmetric.

Turning now to the antisymmetric operator α we can see from its definition (29) that it has a very simple meaning. In fact, let $\mathbf{R}$ be the vector operated upon. Then, by the second of (29), whatever the value of $\mathbf{R}$, $\alpha\mathbf{R}$ is a vector normal to $\mathbf{R}$. Now, this condition can be satisfied by putting

$$\alpha\mathbf{R} = \mathrm{V}\mathbf{w}\mathbf{R},$$

where $\mathbf{w}$ is some fixed vector. But such being the case, we have also, for any $\mathbf{A}$, $\mathbf{B}$,

$$\mathbf{A}\alpha\mathbf{B} = \mathbf{A}\mathrm{V}\mathbf{w}\mathbf{B} = -\mathbf{B}\mathrm{V}\mathbf{w}\mathbf{A} = -\mathbf{B}\alpha\mathbf{A},$$

so that the general definition (29) is completely satisfied. Thus, the antisymmetric operator is, dropping the arbitrary operand,

$$\alpha = \mathrm{V}\mathbf{w};$$

in words, to operate with α is to multiply vectorially by a certain vector $\mathbf{w}$.

Ultimately, therefore, we can write, instead of (31), for any linear vector operator,

$$\varpi = \omega + V\mathbf{w}\,; \tag{32}$$

the symmetric operator ω is half the sum of ϖ and of its conjugate ϖ', while $\mathbf{w}$ is a certain vector characterizing the given operator ϖ. Notice that ω, being symmetric, implies 6 independent scalar data, and $\mathbf{w}$, being an ordinary vector, 3 more, making in all 9, as before. Obviously, ω and $\mathbf{w}$ can be prescribed independently of one another, and these two data (equivalent to $6+3=9$ scalar ones) fully determine the asymmetric operator ϖ.

If we desire to express $\mathbf{w}$ in terms of the coefficients ϖ_{ab}, etc., we can easily do so. For, from the table or the "matrix" (29*a*) we see that

$$\alpha\mathbf{R} = \mathbf{a}(\alpha_{ab}R_b + \alpha_{ac}R_c) + \text{etc.},$$

and since $\alpha_{ab} = \frac{1}{2}(\varpi_{ab} - \varpi_{ba})$, and so on, while $\alpha\mathbf{R} = V\mathbf{w}\mathbf{R}$, we find without difficulty that, if $\mathbf{a}$, $\mathbf{b}$, $\mathbf{c}$ be a *right*-handed system,

$$2\mathbf{w} = \mathbf{a}(\varpi_{cb} - \varpi_{bc}) + \mathbf{b}(\varpi_{ac} - \varpi_{ca}) + \mathbf{c}(\varpi_{ba} - \varpi_{ab}), \tag{33}$$

which is the required expansion of $\mathbf{w}$.

Having thus shown that the antisymmetric part of any operator ϖ is simply a vectorial multiplier $V\mathbf{w}$, it will henceforth be enough to study the remaining part of ϖ, that is to say, the *symmetrical* operator ω.

Principal axes of ω.—Let $\mathbf{R}$ be the operand. Then the vector $\mathbf{R}' = \omega\mathbf{R}$ will in general differ from $\mathbf{R}$ not only in size but in direction as well. But if $\mathbf{R}$ assumes certain particular directions, then it may happen that $\mathbf{R}'$ coincides with $\mathbf{R}$ *in direction*, if not in size. Let $\mathbf{x}$, an unit vector, represent such a privileged direction. Then, $\omega\mathbf{R}$ being a linear function, the inverse direction $-\mathbf{x}$ will, obviously, partake of the same privilege. Such particular directions $\pm\mathbf{x}$ are called *principal axes* of the symmetrical operator ω, both $+\mathbf{x}$ and $-\mathbf{x}$ counting for *one* axis.

This is merely a definition. Let us now see whether at all such axes and how many of them do exist, and what are their mutual relations. Let us start with the last question. Suppose then that there are two different principal axes $\mathbf{x}$ and $\mathbf{y}$. Then, by the very definition of such axes,

$$\left.\begin{aligned} \omega\mathbf{x} &= \omega_1\mathbf{x}, \\ \omega\mathbf{y} &= \omega_2\mathbf{y}, \end{aligned}\right\} \tag{34}$$

where ω_1, ω_2 are some ordinary scalar numbers, which are called the *principal values* of ω, corresponding to these axes $\mathbf{x}$, $\mathbf{y}$. Multiply the first equation scalarly by $\mathbf{y}$, and the second by $\mathbf{x}$, and subtract them from one another. The result will be

$$(\omega_1 - \omega_2)\mathbf{xy} = \mathbf{y}\omega\mathbf{x} - \mathbf{x}\omega\mathbf{y}.$$

But, the operator ω being symmetrical, $\mathbf{x}\omega\mathbf{y} = \mathbf{y}\omega\mathbf{x}$. Thus

$$(\omega_1 - \omega_2)\mathbf{xy} = 0,$$

and if $\omega_1 \neq \omega_2$, we have $\mathbf{xy} = 0$, that is to say, $\mathbf{x} \perp \mathbf{y}$. And should it happen that $\omega_1 = \omega_2$, *i.e.*

$$\omega\mathbf{x} = \omega_1\mathbf{x}, \quad \omega\mathbf{y} = \omega_1\mathbf{y},$$

then λ, μ being any two scalar numbers,

$$\omega(\lambda\mathbf{x} + \mu\mathbf{y}) = \omega_1(\lambda\mathbf{x} + \mu\mathbf{y}).$$

But $\lambda\mathbf{x} + \mu\mathbf{y}$ is *any* vector in the plane $\mathbf{x}$, $\mathbf{y}$. We thus see that if there are two principal axes $\mathbf{x}$, $\mathbf{y}$ to which correspond different principal values ω_1, ω_2, these axes must be *normal* to one another, And if $\omega_1 = \omega_2$, then *every* direction in the plane $\mathbf{x}$, $\mathbf{y}$ is also a principal axis.

Suppose now there is still a third principal axis $\mathbf{z}$ not coplanar with $\mathbf{x}$, $\mathbf{y}$, and let ω_3 be its corresponding principal value, so that

$$\omega\mathbf{z} = \omega_3\mathbf{z}.$$

Then, reasoning as before, we shall see that if ω_3, ω_1, ω_2 are all different, $\mathbf{z}$ will be normal to $\mathbf{x}$ and also to $\mathbf{y}$. And if $\omega_1 = \omega_2 = \omega_3$, then *every* direction whatever will be a principal axis with the same principal value, in which case the operator ω degenerates into an ordinary scalar factor.

Thus, in the most general case the symmetrical operator ω can have *three* different,* mutually perpendicular principal axes, $\mathbf{x}$, $\mathbf{y}$, $\mathbf{z}$; and only three. Because the fourth, if it existed and carried a new ω_4, would have to be normal to those three which, in our space, is nonsense; and if ω_4 were equal to ω_1, say, then the whole plane passing through the fourth and the first axis would consist of principal axes, and since this plane would cut the $\mathbf{y}$, $\mathbf{z}$ plane, ω_2 and ω_3 could not be different from one another, against the assumption.

* *I.e.* such to which correspond different principal values.

Having thus settled the question about the number of the possible different principal axes of ω and their mutual orientation, it remains to see whether they exist, or better, to find them. The technical side of the latter problem will depend upon the manner how ω is given. Suppose it is given through its six different coefficients

$$\omega_{aa}, \quad \omega_{bb}, \quad \omega_{cc}; \quad \omega_{ab}, \quad \omega_{bc}, \quad \omega_{ca}$$

with respect to some arbitrarily fixed framework of normal unit vectors $\mathbf{a}$, $\mathbf{b}$, $\mathbf{c}$, or—which is the same thing—that the three vectors

$$\omega\mathbf{a}, \quad \omega\mathbf{b}, \quad \omega\mathbf{c}$$

are given, say, equal to $\mathbf{A}$, $\mathbf{B}$, $\mathbf{C}$, respectively, so that (ω being symmetrical) $\mathbf{Ab}=\mathbf{Ba}$, etc. Let $\mathbf{x}$ be a principal axis and n the corresponding principal value (both to be found). Then if x_1, x_2, x_3 are the direction cosines of $\mathbf{x}$ with respect to $\mathbf{a}$, $\mathbf{b}$, $\mathbf{c}$, so that

$$\mathbf{x}=x_1\mathbf{a}+x_2\mathbf{b}+x_3\mathbf{c},$$

we have

$$\omega\mathbf{x}=x_1\omega\mathbf{a}+x_2\omega\mathbf{b}+x_3\omega\mathbf{c}=x_1\mathbf{A}+x_2\mathbf{B}+x_3\mathbf{C},$$

and since $\omega\mathbf{x}=n\mathbf{x}$,

$$x_1\mathbf{A}+x_2\mathbf{B}+x_3\mathbf{C}=n(x_1\mathbf{a}+x_2\mathbf{b}+x_3\mathbf{c})$$

or

$$x_1(\mathbf{A}-n\mathbf{a})+x_2(\mathbf{B}-n\mathbf{b})+x_3(\mathbf{C}-n\mathbf{c})=0. \tag{35}$$

From this equation we see that the three vectors $\mathbf{A}-n\mathbf{a}$, etc., are coplanar, so that the volume of the parallelepipedon constructed upon them is nil, *i.e.*

$$(\mathbf{A}-n\mathbf{a})\mathrm{V}(\mathbf{B}-n\mathbf{b})(\mathbf{C}-n\mathbf{c})=0. \tag{36}$$

Since $\mathbf{A}$, $\mathbf{B}$, $\mathbf{C}$ are given, this is a *cubic* equation for the unknown n. Multiply it out and remember that $\mathbf{a}=\mathrm{V}\mathbf{bc}$, and therefore $\mathbf{a}\mathrm{V}\mathbf{bc}=1$. Then the result will be

$$n^3-n^2(\mathbf{Aa}+\mathbf{Bb}+\mathbf{Cc})+n(\mathbf{a}\mathrm{V}\mathbf{BC}+\mathbf{b}\mathrm{V}\mathbf{CA}+\mathbf{c}\mathrm{V}\mathbf{AB})-\mathbf{A}\mathrm{V}\mathbf{BC}=0. \tag{36a}$$

Each of the coefficients of this cubic equation for the principal values of the operator ω has a simple geometric meaning: the first is the sum of the projections of the vectors $\mathbf{A}=\omega\mathbf{a}$, etc., upon the conventional $\mathbf{a}$, $\mathbf{b}$, $\mathbf{c}$, the second the sum of the volumes of the parallepipeda $\mathbf{a}$, $\mathbf{B}$, $\mathbf{C}$, etc., and the last is the volume of the parallelepipedon $\mathbf{A}$, $\mathbf{B}$, $\mathbf{C}$. At the same time we see that these three expressions are *invariants* of ω, *i.e.* independent of the choice of the reference system $\mathbf{a}$, $\mathbf{b}$, $\mathbf{c}$. In fact, if n_1, n_2, n_3 be the principal values of ω, which manifestly are intrinsic properties of the operator, independent of the reference framework, we have, by (36a),

$$\left.\begin{aligned} \mathbf{Aa}+\mathbf{Bb}+\mathbf{Cc} &= n_1+n_2+n_3, \\ \mathbf{a}V\mathbf{BC}+\mathbf{b}V\mathbf{CA}+\mathbf{c}V\mathbf{AB} &= n_2n_3+n_3n_1+n_1n_2, \\ \mathbf{A}V\mathbf{BC} &= n_1n_2n_3, \text{ where } \mathbf{A}=\omega\mathbf{a}, \text{ etc.} \end{aligned}\right\} \qquad (37)$$

These are very important formulae, exhibiting the three invariants of the symmetrical operator ω.

Now, if only $\mathbf{A}, \mathbf{B}, \mathbf{C}$ are real, as we assume, all these invariants, *i.e.* the coefficients of the cubic (36a) are real. That equation has, therefore, at least one real root. Let this be n_1, and let us take the corresponding principal axis * as our reference axis $\mathbf{a}$. Then

$$\mathbf{A}=\omega\mathbf{a}=n_1\mathbf{a}\,; \quad \therefore\ \mathbf{b}V\mathbf{CA}=n_1\mathbf{Cc}, \quad \mathbf{c}V\mathbf{AB}=n_1\mathbf{Bb},$$

and the left-hand member of (36a) becomes at once

$$n^3-n^2n_1+(nn_1-n^2)(\mathbf{Bb}+\mathbf{Cc})+(n-n_1)\mathbf{a}V\mathbf{BC},$$

which is, as it should be, divisible by $n-n_1$, leaving for the remaining two principal values n_2, n_3 the quadratic

$$n^2-n(\mathbf{Bb}+\mathbf{Cc})+\mathbf{a}V\mathbf{BC}=0,$$

which gives

$$\begin{matrix} n_2 \\ n_3 \end{matrix} = \tfrac{1}{2}(\mathbf{Bb}+\mathbf{Cc}) \pm \sqrt{\tfrac{1}{4}(\mathbf{Bb}+\mathbf{Cc})^2-\mathbf{a}V\mathbf{BC}}, \qquad (38a)$$

or, in terms of the coefficients $\omega_{bb}=\mathbf{bB}$, etc., since

$$\mathbf{a}V\mathbf{BC}=\omega_{bb}\omega_{cc}-\omega_{bc}^2,$$

$$\begin{matrix} n_2 \\ n_3 \end{matrix} = \tfrac{1}{2}(\omega_{bb}+\omega_{cc}) \pm \sqrt{\tfrac{1}{4}(\omega_{bb}-\omega_{cc})^2+\omega_{bc}^2}, \qquad (38)$$

so that, if only all the coefficients $\omega_{\iota\kappa}$ are real, these two principal values and, therefore, also the corresponding principal axes are real. That they form with the first axis a normal system we already know.

We have written down the two roots (38) in the assumption that ω was given by prescribing its coefficients $\omega_{\iota\kappa}$ or the vectors $\mathbf{A}$, $\mathbf{B}$, $\mathbf{C}$, with respect to an arbitrary framework $\mathbf{a}$, $\mathbf{b}$, $\mathbf{c}$. But, as a matter of fact, this expansion of the roots is superfluous. For, having taken $\mathbf{a}$ as one of the principal axes of ω, we know beforehand that $\mathbf{b}$, $\mathbf{c}$ will be its remaining two axes, *i.e.* that

$$\mathbf{B}=\omega\mathbf{b}=n_2\mathbf{b}, \quad \text{and} \quad \mathbf{C}=n_3\mathbf{c}.$$

* Whose direction cosines with respect to any $\mathbf{a}$, $\mathbf{b}$, $\mathbf{c}$ might at once be determined from (35) by taking in it $n=n_1$.

Now, with these values, we have $\mathbf{aVBC} = n_2 n_3 \mathbf{aVbc} = n_2 n_3$, so that (38*a*) becomes

$$\begin{matrix} n_2 \\ n_3 \end{matrix} = \tfrac{1}{2}(n_2 + n_3) \pm \sqrt{\tfrac{1}{4}(n_2 - n_3)^2},$$

which is, as it should be, an identity. Thus, the only necessary thing was to state that the cubic (36*a*) has at least one real root, and this was immediately clear.

Having thus ascertained the general properties of the principal axes of ω, let us take them as our (natural) reference system $\mathbf{a}$, $\mathbf{b}$, $\mathbf{c}$, which we will now call $\mathbf{i}$, $\mathbf{j}$, $\mathbf{k}$. Then, n_1, n_2, n_3 being the corresponding principal values, the most general symmetrical linear vector function will be

$$\omega\mathbf{R} = n_1\mathbf{i}(\mathbf{iR}) + n_2\mathbf{j}(\mathbf{jR}) + n_3\mathbf{k}(\mathbf{kR}),$$

that is, n_1 times the first component of $\mathbf{R}$ along $\mathbf{i}$ plus, etc., or using the dot, instead of brackets, as separator,

$$\omega\mathbf{R} = n_1\mathbf{i} \, . \, \mathbf{iR} + n_2\,\mathbf{j} \, . \, \mathbf{jR} + n_3\mathbf{k} \, . \, \mathbf{kR},$$

or dropping the operand $\mathbf{R}$, which it is useless to repeat so many times,

$$\omega = n_1\mathbf{i} \, . \, \mathbf{i} + n_2\,\mathbf{j} \, . \, \mathbf{j} + n_3\mathbf{k} \, . \, \mathbf{k}. \qquad (39)$$

Thus the symmetrical operator assumes the form of what is called (after Gibbs) a *dyadic*, which is a polynomial, in our case a trinomial, of dyads such as $n_1\mathbf{i} \, . \, \mathbf{i}$, etc. It will be well to say a few words on these useful mathematical beings.

Dyads and Dyadics.—The dyads appearing in (39), which, apart of the scalar factors (calling for no explanations), are of the form $\mathbf{i} \, . \, \mathbf{i}$, are but special cases of the general dyad which is

$$\mathbf{a} \, . \, \mathbf{b},$$

$\mathbf{a}$, $\mathbf{b}$ being any two vectors, in general that is not coinciding in direction; the first vector is called the *antecedent*, and the second, the *consequent* of the dyad. The dyad as an operator can be used either as a *prefactor* of the operand, say

$$\mathbf{a} \, . \, \mathbf{bR}, \text{ meaning } \mathbf{a}(\mathbf{bR}),$$

or also as a *postfactor*,

$$\mathbf{Ra} \, . \, \mathbf{b}, \text{ which means } (\mathbf{Ra})\,\mathbf{b}.$$

As Heaviside says somewhere: "A cart may either be pulled or pushed." In fact, this two-fold possibility of attelage of the operator turns out to be very advantageous.

If $\mathbf{a}$, $\mathbf{b}$ are not collinear, then $\mathbf{a}\,.\,\mathbf{bR}$ is, of course, altogether different from $\mathbf{Ra}\,.\,\mathbf{b}$. Such is the case with the most general (asymmetric) dyad. But if the antecedent and consequent happen to be collinear, as in the case of (39), then the dyad, applied to any vector, yields the same result whether it acts as a pre- or a post-factor. Such dyads are called *symmetrical dyads*. They are all of the form

$$\sigma\mathbf{a}\,.\,\mathbf{a},$$

where $\mathbf{a}$ is an unit vector, and σ a scalar. Sums of such dyads are called *symmetrical dyadics*. Thus, the most general symmetrical or self-conjugate linear vector operator $\boldsymbol{\omega}$ may be represented as a (trinomial) symmetrical dyadic. Such, in fact, is (39).

Consider any, generally asymmetric, dyadic, say a trinomial one,

$$\phi=\mathbf{a}\,.\,\mathbf{x}+\mathbf{b}\,.\,\mathbf{y}+\mathbf{c}\,.\,\mathbf{z},$$

which is a certain linear operator. Interchange the antecedents and the consequents; then the resulting operator or dyadic

$$\phi'=\mathbf{x}\,.\,\mathbf{a}+\mathbf{y}\,.\,\mathbf{b}+\mathbf{z}\,.\,\mathbf{c}$$

is called the *conjugate* of ϕ. For such it is according to the previous definition. In fact, if $\mathbf{R}$, $\mathbf{S}$ be any two vectors, we have

$$\mathbf{Ra}\,.\,\mathbf{xS}=\mathbf{Sx}\,.\,\mathbf{aR}, \text{ etc.},$$

since the scalar product is commutative. Thus also the product of $\mathbf{R}$ into $\phi\mathbf{S}$ is seen to be identical with that of $\mathbf{S}$ into $\phi'\mathbf{R}$,—as in the definition (28). Again the product of $\mathbf{R}\phi$ into $\mathbf{S}$ is identical with that of $\mathbf{R}$ into $\phi\mathbf{S}$. Thus no brackets or other separators are needed, and the last property can be written simply

$$\mathbf{R}\phi\mathbf{S}=\mathbf{S}\phi'\mathbf{R},$$

valid for any dyadic ϕ, and its conjugate ϕ'.

We have already seen that the self-conjugate linear operator $\boldsymbol{\omega}$ can be represented as a symmetrical dyadic. We may still mention that the general linear vector operator ϖ can always be reduced to what is called a *normal* (trinomial) dyadic, *i.e.*

$$\varpi=\sigma_1\mathbf{l}\,.\,\mathbf{i}+\sigma_2\mathbf{m}\,.\,\mathbf{j}+\sigma_3\mathbf{n}\,.\,\mathbf{k}, \tag{40}$$

where σ_1, σ_2, σ_3 are scalars (either all positive or all negative), and both the antecedents $\mathbf{l}$, $\mathbf{m}$, $\mathbf{n}$ and the consequents $\mathbf{i}$, $\mathbf{j}$, $\mathbf{k}$ form normal, say right-handed, systems of unit vectors. If these are distinct from one another, we have an asymmetric operator ϖ, and if they coincide, we have a symmetric operator ω, as before. The conjugate of the general ϖ will be

$$\varpi' = \sigma_1 \mathbf{i} \,.\, \mathbf{l} + \sigma_2 \mathbf{j} \,.\, \mathbf{m} + \sigma_3 \mathbf{k} \,.\, \mathbf{n}. \tag{41}$$

The special symmetric dyadic

$$\iota = \mathbf{i} \,.\, \mathbf{i} + \mathbf{j} \,.\, \mathbf{j} + \mathbf{k} \,.\, \mathbf{k}$$

leaves, of course, any vector operand $\mathbf{R}$ intact, and is, therefore, called an *idemfactor*. It is also, for all purposes, equivalent to 1. And if σ be any scalar, then $\sigma\iota$ as an operator is equivalent to σ itself, as an ordinary numerical factor. Thus, expressions such as $\sigma + \mathbf{a} \,.\, \mathbf{b}$ will again be dyadics, and require no further explanations.

To close this section it will be enough to make a few remarks on the "multiplication," *i.e.* the successive application of dyadics.

If $\phi = \mathbf{a} \,.\, \mathbf{b}$ and $\psi = \mathbf{c} \,.\, \mathbf{d}$ be two dyads, and $\mathbf{R}$ any vector operand, we have obviously

$$\phi(\psi\mathbf{R}) = (\phi\psi)\mathbf{R},$$

where $\phi\psi$ is the dyad $\mathbf{a}(\mathbf{bc}) \,.\, \mathbf{d} = (\mathbf{bc})\mathbf{a} \,.\, \mathbf{d}$. Similarly, if γ be a third dyad, we have

$$\phi\psi(\gamma\mathbf{R}) = \phi(\psi\gamma)\mathbf{R} = (\phi\psi)\gamma\mathbf{R},$$

the *associative* property, so that each of these expressions can be simply written $\phi\psi\gamma\mathbf{R}$. And the same is easily seen to hold if ϕ, ψ, etc., stand for binomial or polynomial dyadics. Again, since the scalar product of vectors is distributive, we have for any dyadic ϕ and any vectors $\mathbf{R}$, $\mathbf{S}$,

$$\phi(\mathbf{R} + \mathbf{S}) = \phi\mathbf{R} + \phi\mathbf{S},$$

and also, if ψ, γ be two more dyadics, the operational equation

$$\phi(\psi + \gamma) = \phi\psi + \phi\gamma,$$

and also

$$(\psi + \gamma)\phi = \psi\phi + \gamma\phi.$$

In short, the *distributive* property holds for the multiplication of any polynomials of dyads, and therefore of dyadics. Such products can, therefore, be expanded as in ordinary algebra, the only necessary precaution being to keep the order of the operators and of the constituents of the dyads intact, since (in general) the commutative property does not hold. Thus, for instance,

$$(\mathbf{a} \,.\, \mathbf{b} + \mathbf{c} \,.\, \mathbf{d})(\mathbf{e} \,.\, \mathbf{f} + \mathbf{g} \,.\, \mathbf{h}) = (\mathbf{be})\mathbf{a} \,.\, \mathbf{f} + (\mathbf{bg})\mathbf{a} \,.\, \mathbf{h} + (\mathbf{de})\mathbf{c} \,.\, \mathbf{f} + (\mathbf{dg})\mathbf{c} \,.\, \mathbf{h}.$$

Vectors not separated by dots are fused into scalar products, as (**be**), (**bg**), etc., and here of course the order is irrelevant; but it must be carefully preserved in the resulting dyads, such as $\mathbf{a}.\mathbf{f}$, not $\mathbf{f}.\mathbf{a}$ (unless **a**, **f** are collinear). Apart from this precaution, the multiplication of dyadics is as easy and convenient as the common multiplication of polynomials, and it will be found to render inestimable services in the treatment of many geometrical and physical, especially optical, problems. Some illustrations of the latter kind will be found in the "Simplified Method, etc.," mentioned before. The final result of such multiplications of two or more polynomials will be a polynomial of dyads, say

$$\mathbf{A}.\mathbf{B}+\mathbf{C}.\mathbf{D}+\mathbf{E}.\mathbf{F}+\mathbf{G}.\mathbf{H}+\text{etc.};$$

but since each of these antecedents and consequents can be expressed in the form $x\mathbf{a}+y\mathbf{b}+z\mathbf{c}$, where **a**, **b**, **c** are any non-coplanar unit vectors, any such result can, in the first place, be reduced to a sum of nine dyads, viz.

$$\sigma_{11}\mathbf{a}.\mathbf{a}+\sigma_{22}\mathbf{b}.\mathbf{b}+\sigma_{33}\mathbf{c}.\mathbf{c}+\sigma_{23}\mathbf{b}.\mathbf{c}+\sigma_{32}\mathbf{c}.\mathbf{b}+\ldots+\sigma_{21}\mathbf{b}.\mathbf{a},$$

and it can be proved that this can always be reduced by a proper choice of two orthogonal systems, **i**, **j**, **k** and **l**, **m**, **n**, to the *normal* form (40), which is that of any, generally asymmetric, linear operator ϖ. Ultimately, the latter can with advantage be split into a symmetric operator ω and the simple operator $\mathbf{Vw}$, as in (32).

10. Hints on Differentiation of Vectors. The concepts of differentiation and integration as applied to vectors do not belong to the subject proper of this booklet, which is vector Algebra. Yet a few elementary remarks on the differentiation of vectorial expressions may be here added, as they are likely to be useful to some readers, and as they do by no means require much space.

Let **R** be a variable vector. To have a possibly desirable picture think of **R** as the position-vector of a particle moving about in space, round a fixed origin. Let t be any independent scalar variable, say the time. Then, $\Delta\mathbf{R}$ being the vector increment, *i.e.* the vector drawn from the position of the particle at the instant t to that at a later date $t+\Delta t$, the quotient $\Delta\mathbf{R}/\Delta t$ will be a certain vector, having a definite tensor (size) and a definite direction.

We may call it provisionally the average vector-velocity of the particle. If this quotient (a vector) tends, with indefinitely decreasing Δt, to some definite limit, definite both in size and direction, we call this limit-vector the ***derivative*** or the fluxion of $\mathbf{R}$ with respect to t (or the vector velocity of the particle), and denote it by $\frac{d\mathbf{R}}{dt}$ or $\dot{\mathbf{R}}$. In short symbols

$$\frac{d\mathbf{R}}{dt} = \dot{\mathbf{R}} = \text{Lim}\,\frac{\Delta\mathbf{R}}{\Delta t}.$$

This vector will, in our illustration, be tangential to the orbit of the particle, and its tensor will represent the particle's speed.

From this definition it follows at once that

$$\frac{d}{dt}(\mathbf{R}+\mathbf{S}) = \frac{d\mathbf{R}}{dt} + \frac{d\mathbf{S}}{dt} = \dot{\mathbf{R}} + \dot{\mathbf{S}},$$

where $\mathbf{R}$, $\mathbf{S}$ are any vector functions of the variable t. And, if $\mathbf{r}$ be the unit of $\mathbf{R}$, so that $\mathbf{R} = R\mathbf{r}$, we have of course

$$\dot{\mathbf{R}} = \dot{R}\mathbf{r} + R\dot{\mathbf{r}}.$$

Again, since the scalar product of two vectors is distributive, so that $\Delta(\mathbf{RS}) = \mathbf{R}\Delta\mathbf{S} + \mathbf{S}\Delta\mathbf{R}$ *plus* terms of higher order, we have

$$\frac{d}{dt}(\mathbf{RS}) = \mathbf{R}\dot{\mathbf{S}} + \dot{\mathbf{R}}\mathbf{S}.$$

In particular, if $\mathbf{r}$ be a unit vector, so that $\mathbf{r}^2 = 1$, we have, by differentiating the latter condition, $\mathbf{r}\dot{\mathbf{r}} = 0$, so that $\dot{\mathbf{r}} \perp \mathbf{r}$, which is also an obvious property. Similarly, for the vector product, which again is distributive,

$$\frac{d}{dt}\mathbf{VRS} = \mathbf{V}\dot{\mathbf{R}}\mathbf{S} + \mathbf{VR}\dot{\mathbf{S}},$$

the only precaution being to preserve the order of the factors, or —if this be inverted—to change the sign of the product in question. In quite the same way we have

$$\frac{d}{dt}\mathbf{AVBC} = \dot{\mathbf{A}}\mathbf{VBC} + \mathbf{AV}\dot{\mathbf{B}}\mathbf{C} + \mathbf{AVB}\dot{\mathbf{C}},$$

and so forth.

Even the case of linear vector functions, such as $\varpi\mathbf{R}$, does not call for lengthy explanations. If not only the operand $\mathbf{R}$, but also the nature of the operator ϖ varies with t, we have

$$\frac{d}{dt}(\varpi\mathbf{R}) = \varpi\dot{\mathbf{R}} + \dot{\varpi}\mathbf{R},$$

since ϖ is distributive. Here $\dot{\varpi}$ is the derivative of the operator. If, for instance, ϖ is represented as a dyadic, say

$$\mathbf{A}\,.\,\mathbf{B}+\mathbf{C}\,.\,\mathbf{D}+\mathbf{E}\,.\,\mathbf{F},$$

we have

$$\dot{\varpi}=\dot{\mathbf{A}}\,.\,\mathbf{B}+\mathbf{A}\,.\,\dot{\mathbf{B}}+\ldots+\mathbf{E}\,.\,\dot{\mathbf{F}}.$$

And if the form $\omega+\mathrm{V}\mathbf{w}$ is used, we have

$$\dot{\varpi}=\dot{\omega}+\mathrm{V}\dot{\mathbf{w}},$$

where $\dot{\omega}$ can again be expanded, as the derivative of a symmetrical dyadic. It is scarcely necessary to add any further explanations.

INDEX

Addition of vectors, 3-7
algebraic sum, 7
antecedent, of dyad, 35
antisymmetrical operators, 29
area, of parallelogram, 17
associativity, of addition, 6
—— of dyadics. 37
asymmetrical operators, 29
autoproduct, scalar, 13
axes, of operator, 31-35

Chain of vectors, 3
closed chains, 6
coinitial vectors, 3
collinear ——, 7
commutativity of addition, 6
—— of scalar product, 12
components, 8
conjugate operators, 29
consequent, of dyad, 35
constituents, of vector operator, 28
continuous operators, 26
coplanar vectors, 20
cosine formula, 16

Derivative of vector, 39
—— of operator, 40
determinantal form of vector product, 22
difference of vectors, 10
differentiation of vectors, 39
distributivity, of dyadics, 37
—— of linear vector operators, 26
—— of scalar product, 14
—— of vector product, 20
dyads, and dyadics, 35-38

Equality of vectors, defined, 2-3

Free vectors, 2
function, vector-, 26

Gibbs, 35

Heaviside, 35

Idemfactor, 37
invariants, of operator, 33
iterated multiplication, 23-25

Linear vector operator, 27
localized vectors, 2

Multiple of a vector, 7
multiplication of dyadics, 37-38

Negative factor, 7
nil vector, 6
normal and longitudinal parts of a vector, 36
normal form of dyadic, 36

Operators, 26
origin, of vector, 1

Parallelogram, and vector sum, 5
polar coordinates, 9
position vector, 25
postfactor, 35
prefactor, 35
principal axes of operator, 31
—— values ——, 32
product of vectors, scalar, 12
—— vectorial, 17
projection, 14
Pythagoras' theorem, 15

Reference system, 9
reflector, 26
refraction, 26
right-handed system, 18

Scalars, 1
scalar product of vectors, 11-13
self-conjugate operators, 28
separators, 14
sine formula, 23
skew operators, 29
spherical trigonometry, 16, 23
square of a vector, 13
stretcher, 26
subtraction of vectors, 10-11
sum of vectors, 3
symmetrical dyads, 36
—— operators, 28

Tensor, 1
translation, 3

Unit vectors, 1, 8

Values, principal, of operator, 32
vector, defined, 1
vector product, defined, 17
volume, of parallelepipedon, 19, 22

Printed in Great Britain
by Amazon